AF846230

Religion and Agriculture:
Sustainability in Christianity and Buddhism

Lindsay Falvey
2005

IID

Institute for International Development
Adelaide

Publication of this work was assisted by a publication grant from the University of Melbourne.

Cover Design: Christopher John Falvey

This edition published by:
The Institute for International Development,
14 Birdwood Street,
Netherby,
Adelaide, South Australia 5062
Australia
<http://www.iid.org>

National Library of Australia Cataloguing-in-Publication entry:
Falvey, J. Lindsay.
Religion and agriculture : sustainability in Christianity and Buddhism.
Bibliography.
ISBN 0 9751000 2 5.
1. Sustainable development - Religious aspects - Buddhism.
2. Sustainable development - Religious aspects - Christianity.
3. Agriculture - Religious aspects - Buddhism.
4. Agriculture - Religious aspects - Christianity.
I. Institute for International Development.
II. Title.
338.927

Copyright © 2005 Lindsay Falvey

Dedicated to

The Reverend Dr. A. Harold Wood OBE,

whose theological tutoring
for the Methodist Lay Preachers Course
spurred a lifelong interest
in the meaning of scriptures.

Books by the Same Author

Karn Kaset Thai:
U Khao U Nam Kham Sahasawat. 476pp. (2005) – [Thai language]

Sustainability: Elusive or Illusion?
Wise Environmental Intervention. 245pp. (2004)

Buddhism Briefly Explained. 50pp. (with Siladasa). (2004)

The Buddha's Gospel:
A Buddhist Interpretation of Jesus' Words. 108pp. (2002)

Thai Agriculture:
Golden Cradle of Millennia. 490pp. (2000)

Smallholder Dairying in the Tropics
(with Charan Chantalakhana). 447pp. (1999)

Land and Food: Agricultural and Related Education in the Victorian Colleges and the University of Melbourne (with Barrie Bardsley). 266pp. (1997)

Food Environmental Education:
Agricultural Education in Natural Resource Management. 280pp. (1996)

International Consulting: Providing Services to International Development Agencies. 185pp. (1994)

Introduction to Working Animals. 198pp. (1988) – [also published in Vietnamese language]

Cattle and Sheep in Northern Thailand. 104pp. (1979)

IID
The Institute for International Development

This book is published by the Institute for International Development Ltd, an Australian not-for-profit public company limited by guarantee, which through its Institute for International Development Fund supports publications; seminars; workshops; international exchanges; research; education, and culture.

Professor Lindsay Falvey is currently Visiting Fellow, Clare Hall, University of Cambridge. He was previously Chair of Agriculture and Dean of Land and Food at the University of Melbourne. He holds degrees from Latrobe, Melbourne and Queensland Universities, including two doctoral degrees, and is the author of some ten books and more than 100 professional papers on themes relating to society, food production and international development. Three of his recent works relate to religion and wise agricultural practice on a global basis. He has also managed and consulted for international development companies working in some 20 countries, particularly in Asia and especially in Thailand. He was elected as a Fellow of the Academy of Technological Sciences and Engineering in 1997 and of his profession's institute in 1991, and has received several national and international awards for his contributions to international and rural development, including the Australian Centennial Medal in 2003.

Contents

Acknowledgements

Acknowledgements are usually restricted to the persons and financiers of the work they preface. This is important and I will come to that shortly. In this case, inspired by a thought from a seminar at the University of Cambridge while completing this book, I wish to acknowledge where I am coming from. In Buddhist terms, this might be seen as causes and conditions. The idea from the seminar immediately made sense to me, as I usually read about authors and where they are coming from before making a decision to read a book. So, here the acknowledgements begin with a perhaps more objective description of the author than may be found on the book's cover.

I have been influenced by Protestant theological training as a young man, thirty years of association with Asia including living in Thailand for five years, a short period in a Thai temple, and by following some Buddhist practices. Inquisitive and sceptical, I have read eclectically and am more impressed by testable and enduring linkages across disparate subjects than by received facts. Perhaps this last point explains why my professional scientific interests have moved progressively from technical to theoretical to esoteric as my 'career' progressed. I should also acknowledge that this book began as a personal search for understanding about the yawning gap between scientific and religious understanding of sustainability in the light of obvious yet neglected experience in Asia. With that acknowledgement of my intellectual origins, I would like to thank various persons who have provided towards the book's success.

It has been several years since I began studying the ideas on which this book is based, and over those years many have contributed order and analysis to sometimes confused and confusing thoughts. I cannot mention them all here, but I hope they will know that I value their influences when they receive their complementary copy of the book.

Among friends and colleagues that I would especially like to thank is Will Johnston who encouraged me in the research and who read the draft manuscript. Also, those who have hosted me including Doug and Lynn Forno whose Cairn's veranda served for collation of a decade's notes, some of which had been made in their earlier Philippines' abode. It also includes the University of Melbourne which provided its usual well resourced research environment. And more recently it includes Clare Hall, Cambridge, through their invitation to live in that magical environment to complete the manuscript. This occurred through the good offices of Ekhard Salje, David Gosling, Joseph Runzo and Nancy Martin, to each of whom I remain most grateful.

I particularly would like to thank those who commented on the draft manuscript. I hope they will find this book a vast improvement on the versions they have read, for the final product incorporates many of their thoughts, criticisms and many new ideas stimulated by those interactions. In particular, I would like to thank Simone Bernhardt, David Gosling, Elizabeth Heij, Will Johnston, John Leake, Tony Loquet, Robyn Porter, Rene Salm and Robert White for both their general and detailed comments. I would also like to thank Gabriel Fragnière, whose enthusiasm after a chance meeting in Strasbourg began a productive and professional relationship that has influenced the book's internal presentation. And for the outward presentation, I am proud to acknowledge the cover design work of Christopher Falvey. Publication of this work was assisted by a publication grant from the University of Melbourne, which I also gratefully acknowledge together with the related efforts of Snow Barlow.

It is at this point that an author usually accepts the remaining errors and omissions with feigned humility. For my part, I can assure readers that errors exist in the broad areas covered. But there may also be points that may seem wrong to a reader, but do not seem so to me. Not at this time in any case. Please bear with them.

Lindsay Falvey
Clare Hall, Cambridge

Introduction

he who'll but on sense rely
reality doth yet deny

Science and religion are natural bedfellows kept apart by the prissy maiden-aunt of modern convention to the detriment of the very culture of both. Nowhere is this more evident than in the long religious history of our self-understanding and the environmental manipulation that we term the science of agriculture. A means of redressing this is sometimes felt to exist in 'sustainability' – though we do not know what it really means.

This book uses sustainability as the meeting point of science and religion. It does this by accepting that human knowledge is contained in spiritual wisdom at least as much as in scientific insight and by using both to examine the elusive subject of sustainability. One purpose of this book is therefore to highlight forgotten human resources that are ever at our disposal. To this end, it brings Eastern and Western insights to the subject of science and sustainability through consideration of different religious teachings. The approach of the book itself borrows from that Oriental discourse which treats a subject from multiple perspectives without dismissing conflicting views. From such an approach, a higher level of understanding can be revealed – perhaps even truth – in which conflicts dissolve into unity. This may not be as radical as it sounds, for when modern science suspends its precious rationality to allow competing theories on the nature of light as particles or waves, for example, it may in fact owe the same debt to Eastern approaches. So from this point onwards, agriculture, religion and sustainability are viewed from different and multiple viewpoints to reveal a common insight.

Why choose agriculture as the entry point to sustainability? Why even relate religion to the search for sustainability? Such obvious first questions of this research are answered in the following chapters in many ways, some of which present profound insights of masters. For the purposes of this introduction, we can accept the usual inference that sustainability is responsible environmental care. Such care would be most effective in our most basic and most significant interference with nature - agriculture. We might also accept that our actions in the name of sustainability include beliefs of responsible approaches and actions to forestall change that share some characteristics with religion - such as immortality and stewardship. It is perhaps more than a curiosity then that agriculture and religion are inextricably joined in our history. For it was agriculture that allowed the settled division of labour which produced the classes that codified our religions, and in explaining their deeper intents, religious writers used metaphor from the main occupation of their day, agriculture.

But then it is usually asked - 'what is sustainability anyhow?' Is it a method, an output, or an attitude? Is it sustaining of the production base for continued food output, or sustaining the essential 'naturalness' of the environment? Or is it sustaining of profits as a basis for ensuring self-interested continuation of care in order to derive future profits? In fact it is all of these and much more in terms of the word's usage. But as this work reveals, sustainability is ultimately an attitude that shares something with religion, and much more with the transcendental messages that organized religion sometimes obscures.

A facile conclusion from this work would be that the less a system interferes with the natural environment the more sustainable it is. This may be so, but it is not useful knowledge unless it is understood that what makes production of food sustainable is not some group of technologies but an attitude that must pervade all aspects of life. To be explicit, sustainability involves challenging our self-awareness not only in our production of food, but in our consumption and all that we do in between. And awareness requires a current knowledge base and an open mind for anyone living in a modern society. To mistake dogmas about 'traditional' agriculture or 'organic' food for sustainable agriculture is as superficial as to mistake religious dogma for its original spiritual intent. Nevertheless, these pages also speak favourably of traditional food production when it is understood in scientific, cultural, spiritual and environmental terms.

The human processes of food production are grossly misunderstood and are seldom clarified by those who benefit from promotion of images of food shortages on the one hand and idyllic peasants on the other. A few facts illustrate the point: most food in the world is produced in Asia, not in the Western world; ninety percent of the world's food is not traded internationally, rather, it is consumed in the country where it is produced; food is not in short supply, in fact current food production exceeds requirements for all six billion of us to eat well if we all had equal access to food; nor need there be future food shortages as known agricultural technologies can provide sufficient food for the projected future world population. When we argue against these facts, or when we consider the environmental degradation of modern agricultural land use to be intolerable, we might do well also to consider that we could feed ourselves from fermentation products that require minimal use of land. But that is mere technology, which of itself has little to do with sustainability – for it is our attitudes that ultimately determine the matter.

Modern Western attitudes are largely unguided by absolute ethics or by insight of the cultures' unseen seers.

These anthropocentric attitudes have produced a utilitarian basis for evaluating sustainability. At the same time, the separation of food production from consumption insulates urban ecological sensitivities from agricultural reality. The divide also seems to restrain the contributions of urban-oriented religion, philosophy and science to rural agriculture. The divide is bridged where science produces financially rewarding technologies such as pesticides – but often at the cost of operating in a vacuum remote from public awareness, and from religion and philosophy. Thus arise partial definitions of ethics according to each party's different interests. But the inherently artificial separation of modern secular and traditional religious ethics is unlikely to persist in the pragmatic field of agriculture. The book brings these together.

So, to inform the sciences of agriculture and sustainability, this work focuses on the dualities that underpin modern and perennial conundrums. It tests such conflicts as those between sustainability and self-interest, between environmental change and apocalypticism, between competition and cooperation, and between atomization and holism. It examines the possibility of a return to non-domineering matriarchal values outside the modern technical paradigm, and separate from the fallacy of sustainability as a cross-generational genetic proxy for immortality. It considers whether linearity in science is akin to a second Fall, whether continued economic growth is necessary, and whether sustainability can exist at all in a consumerist society. In order to approach these diverse dualities, the work often contrasts Western and Eastern attitudes, actions and teachings. But it does not blindly accept generalisations – for the West is no more universally uncaring than the East is universally spiritual, or poor. This approach elicits some general lessons about our fundamental understanding of self and nature.

In general, we can say that we must all understand nature in a different way. We can no longer see it only in the utilitarian terms of 'what it can do for us?' – but more as a dynamic flow to which we continually adapt. That seed is perhaps planted through such characteristically Buddhist-influenced thoughts as care for all things in their natural conditions separate from non-essential benefit to us. As we continually test the limits to our own growth as a species in ways from population to production of food, we are increasingly reliant on wisdom and insight as well as on good science. A dominant selfish anthropomorphism will become more hazardous with time. In this context, most see sustainability as the only future, even if it is only an ideal. And the relationship between insight and sustainability points to a need for each of us to widen our consciousness to recognize and cultivate wisdom.

The book therefore embraces wisdom as well as knowledge. It is an intellectual history, an exercise in cross-religious comparison, an exercise in Western and Oriental cross-cultural understanding, and an exploration of science and religion. It takes multiple approaches as a means of acknowledging the fact that most agricultural scientists are uneducated about religion, and that most religionists and theologians are uneducated in the sciences of agriculture. Of course, some individuals may feel they understand both fields but do any of us really know the science and the religion of a subject? It is extremely difficult to contextualize agriculture and religion at the same time. From a religious point of view, religion and nature including natural resources form a magnetic continuum, while from the point of view of agricultural science, agriculture and nature always seem to attract each other. The former view is normative and the latter practical, yet the common word for each is 'nature' – nature conceived in two quite different ways. To bridge this gap, the following discussion shifts from one context to another. While agriculture is the means by which we will examine sustainability, the method and conclusions of this work apply generally to sustainability – as indeed they apply to our individual relationships with nature which includes ourselves.

Agriculture is a suitable basis for examples of our relationship with nature as it represents our first major interference in the natural environment, one which continues today with more far-reaching tools at our disposal. It is our most widespread modification in nature. As we, the whole human race, sought to understand our place in nature, we discovered a unity that includes us and which is understandable through a marriage of intuition and rationality. In codifying this insight as religion, the examples, metaphors and analogies drawn from agriculture assisted understanding for 'those who had ears to hear'.

From our progressive enhancement of our management of nature we have reached today's amazing facility to transform it, from atmospheric to molecular levels. And in so managing nature, we seek to further enhance or at least sustain the current situation – to hold onto what we have while we seek more. Thus we have directed much of our effort and belief and even more of our rhetoric into 'sustainability' as a virtue – any change not to our liking is wrong. Ironically, this leads us to ignore the cycles of nature.

In any discussion of sustainability, time is critical. Likewise, in any discussion of religion, immortality or rebirth seems to be a critical element. In seeking to maintain an artificial situation to which we are attached and labelling that virtuous act 'sustainability' we seek to stave off the ultimate change of death – to cheat time. Such cheating requires us to reinterpret scriptures to show sustainability as a virtue by using the same misinterpretations as those related to immortality. If our societal angst is a product of such counterintuitive and counter-intellectual falsehoods – false because our lives will end sometime – then we have the seeds to our own growth within ourselves. And those seeds grow into the realisation that our angst, our suffering, is self-inflicted even for events that we did not cause. Just because the sun will one day become too hot or reduce to a 'red giant' or some other non-effective source of energy is not a cause for worry unless we fear oblivion.

Within a shorter time frame, we might interpret sustainability as a steady-state rather than an equilibrium condition and recognize that the system is always in flux. With that flux regular input and output assumptions appear to be valid, genetic diversity is maintained by some species dying out and new ones evolving, food is produced and consumed far from its points of production but within a system that allows sufficient time for natural regeneration. But such a system relies on constant environmental conditions and this has never been the case. If sustainability exists, it would seem that it is a relative rather than an absolute state.

The distinction between relative and absolute sustainability will be readily assimilated by readers of Buddhist philosophy who will see it as a consequence of the conception of relative and absolute truth. Absolute sustainability is unattainable in the world as we know it, but may be attainable in the universe as a whole. On a more mundane level, relative sustainability is attainable and its relativity is dependent on time. If our impacts on an ecosystem are slow, the system may well have time to evolve and adapt to the changes taking place – this is what occurred in much agriculture across the millennia. Of course, this argument, while popular in its conservatism, applies to major changes also, even cosmic cataclysms – the only difference is that we may not evolve or adapt fast enough to survive.

We may well have been in a period of relative sustainability from our first appearance as *homo sapiens* until some 150 years ago if we take the arbitrary point of chemical fertilizer use as a turning point. Since that time, the pace of change has accelerated and we have forced the system into a period of relative un-sustainability. We can then refine our question to ask – can we return to a position of relative sustainability? It seems that the source of the accelerated change, population increase and continued encouragement of greed as a social system, would preclude even that relative balance being restored. To return to the Buddhist concept of relative versus absolute reality, we acknowledge that we can answer logically to all relative questions but this does not produce absolute truths and thus not absolute sustainability. In our un-enlightened existence, absolute sustainability remains but an ideal. We shall be considering both concepts – relative sustainability as wise interactions with nature, and absolute sustainability as an ideal like any other spiritual ideal.

Examining spiritual ideals requires wisdom – that form of understanding we often strenuously eschew in science as unprovable intuition. And in the language of wisdom, if we are not enlightened, how can we know what is wise and what is not? To do justice to sustainability, or agriculture, religion or cross-cultural dialogue for that matter, we do well to abandon the empirical reductionism of usual Western analysis and to embrace a wider approach. An alternative is an 'Asian' approach which often appears to the Western reader to wander around a subject regarding it from all viewpoints while permitting apparently contradictory conclusions. Another approach is the re-marriage of rational and intuitive knowledge in the manner of William Blake who saw the works of his contemporaries such as Bacon, Locke and Newton as neglecting the breadth of human intuition – to him, philosophy and science were but part of human wholeness. It is with a blend of these two approaches that we enter the following pages to examine sustainability, agriculture and religion together, and from all sides. We may then compare our conclusions from each of the perspectives. Each chapter may also be seen as a separate reflection in its own right, with its summary conclusions linking that view to the next.

To begin our voyage around religion, agriculture and sustainability, the first chapter introduces some questions and themes to be examined, and shows how these three usually separated subjects are related and how they cannot be separated without reducing our understanding.

Chapter 1

Seeking Agricultural Sustainability: Science and Spirituality

Religion without science is blind; science without religion is lame.
Einstein

Sustainability has attracted a motherhood cachet that produces unhealthy outcomes. As an adjectival prefix, it is used to render almost any human action above criticism, while debasing its currency and accelerating its circulation in the mode of the bad penny. Within the context of agriculture and resource management, 'sustainability' is widely applied to intensive systems that show no indication of being sustainable. Perhaps such terms as 'sustainable exploitation of natural forests' reveal a Freudian conflict between our assumed right to dominate and our existential guilt. Of course, all living things 'exploit' something else, but in this knowledge, why do we feel that the word is negative? In the science fields, we hypocritically use 'sustainability' to enhance our search for more funds for research, yet at the same time tacitly encourage a blind belief in its attainability among the less informed public. It is difficult to think of any other term in modern science that we have so misused and confused.

With such confusion of definitions, we can do better than lexicographical analysis and transient political dogma, for it is the origin and development of the concept that might explain why this word 'sustainability' has proved so durable in fields where fads routinely vanish quickly. It is perhaps best known in the form of 'sustainable development' that arose from United Nations influence, and which in fact includes acknowledgement of the role of world religions. Yet, the word seems to plumb a primal and anti-intellectual depth that has evolved into a realm of psychology from its earlier location in philosophy and religion. Our preoccupation with 'the environment' might be one explanation for the durability of 'sustainability', but to me it seems to attach to something deeper. I have therefore chosen to examine the agricultural sciences which serve our most pervasive interaction with the terrestrial environment. And the approach I have taken is to examine some cultural underpinnings of agricultural sustainability as expressed in religious and spiritual understandings.

The argument overarches other recent works, which have produced, on the one hand, a consideration of Buddhism and agriculture in Thailand,[1] and on the other, a unification of diverse papers which suggest that our modern use of science limits progress towards sustainability.[2] The gist of these two publications can be appreciated from the following extract.

> *Sustainability of the environment contains both wishful thinking and ignorance – ignorance of the reality that natural systems are complex and unfathomable by scientists, and that repetition of their outputs depends on repetition of initial and all subsequent conditions. ... Ironically, global forces, now incorporating sustainability in their programs, assume local guises that often displace existing agricultural systems in less-developed countries that seem to have been environmentally stable over millennia. ... Two sources of knowledge inform all discussions, rationality including the technological understandings of science, and the insights of spiritual masters. To consider sustainable agriculture within a modern technical paradigm has led us to a perpetually uncertain attempt to sustain an output by constant technological innovation.*

To date, I have adhered to the convention of separating the spiritual aspects of life from the cultural, including the

---

[1] Falvey (2000)
[2] Falvey (2004)

religious, aspects. But that is not the way most of us experience life. So, while remaining faithful to scientific discourse, we do well to examine the gap between religion and spirituality. This is not a new or deep philosophical division. It is simply the consideration of cultural observances and beliefs separate from personal experience and its cultivation. Einstein's insight that *religion without science is blind; science without religion is lame* captures the essence – the true value of religion to humankind requires not only belief but objective rational thought and experience to widen into a spiritual dimension. The real insights of science only occur when that spiritual dimension is engaged.

Obviously spiritual insight has been a source for religion and remains a component of it, albeit often somewhat remote. In the common definition of religion as a culturally-bound belief system, rites and traditions are more important than cultivation of spiritually experiential development or mystical insights. Lest some readers who see little value in either religion or spiritual matters are already worried, let me hasten to emphasize that every society has had and retains some vestige of religion, and virtually all have included beliefs about life after death.[3] As our close relatives the chimpanzees do not appear to share such inclinations, it has been suggested that religion arose simply to foster the group solidarity[4] necessary for humans to attain an evolutionary advantage.[5]

Seen from this perspective, agriculture, which is based on settled communities and is often assumed to have provided the stability needed for cultural development including sophisticated religion, may be seen more as a product than a source of religion. In any case, agriculture and religion seem to have been intertwined since their respective invention.

---

[3] Hull and Bold (1994)

[4] Moore (2000)

[5] Simon (1990)

As the tribes from which we are derived expanded, their religions spread and absorbed new ingredients until the great universal religions arose - religions not limited to a select group. These great religions that have survived until today, though differing from each other, have always contained similar insights, such as has been demonstrated in the Indian Mahayana Buddhism and the Near-eastern Gnosis[6] that also had a major effect on the development of Christianity. In fact the early teachings of such teachers as the Buddha and Jesus may even relate more closely to each other than to their own subsequent traditions.[7]

Do such similarities suggest contact between these emerging cultures? Certainly this is indicated from such observations as: the mutual influence of art of the era; Alexander's protection of Buddhist communities in what is now Pakistan; the sea and land trade routes across the region; the Greek philosopher Pyrrho having travelled to India and having been influenced by Indian insights,[8] and the many other peripatetic scholars who roamed the whole region in search of spiritual knowledge. The rock and cliff edicts[9] of the Indian king Asoka, which date from the third century BCE, document contact with Antiochus II, Theos of Syria, Ptolemy, Antigonas, Magas, and Alexander. Others note that Pythagoras, an initiate of Cretan rites and evidently familiar with Eastern thought, was a contemporary of the Buddha.[10]

What does all this have to do with sustainable agriculture? Quite a lot as we shall see. For example, the supplantation of pre-historical matriarchal religious structures by patriarchal religions may well have accompanied fundamental changes in attitudes to agriculture. This is

---

[6] Conze (1975)
[7] Falvey (2002)
[8] Flintoff (1980)
[9] Hultzsch (1925)
[10] Abraham (1990)

suggested in the Orphic tradition of the early Neolithic period with its fertility and related rites being seen as diffusion from the Indian Dravidians into Sumer, Egypt and Crete to appear as Dionysism prior to the Aryan influence in India. Mythically this is relayed in personal terms of the passionless Orpheus being killed and co-opted by the maidens of Dionysius. Vestiges remain in such forms as the parallels between Orpheus and Christ as shepherds, saviours of souls, and feeding the multitudes on bread and fish.[11] Our search for such havens as sustainable agriculture may be part of our attraction to such enduring and powerful metaphors as shepherds of lost souls and saviours that reunite us with the divine and feed our inner feeling of loss.

Our lost natural state, with its feminine terminology, may be an unconscious yearning expressed, for example, in the desire to return to Eden. That yearning may also be seen when we seek to replace dominance with partnership through myths, psychologically astute teaching devices and the creation of a 'sustainable future'. And in searching for that sustainable future among the deluge of information and exotica of globalization, the West is encountering different worldviews. One of these different views is now labelled Buddhism. Until encountered by the West, it knew itself simply as 'the Dharma', which among its teachings maintains a sceptical attitude to teachings that are not tested in one's own experience. It also logically associates liberation from suffering with such characteristics as compassion and loving-kindness.

The arrival of Buddhism in the West is largely a phenomenon of the last 150 years and possibly began in the rebellion of the romantics against the industrial outcome of the Western enlightenment. One of van Gogh's letters captures this. *If we study Japanese art we see a man who is undoubtedly wise,*

[11] Abraham (1994)

philosophic and intelligent, who spends his time doing what? In studying the distance between the earth and the moon? No. In studying Bismarck's policy? No. He studies a blade of grass. But this blade leads him to draw every plant and then the seasons, the wide aspects of the countryside, then animals, then the human figure. Isn't it almost a true religion which these simple Japanese teach us, who live in nature as though they themselves were flowers?

Some of my colleagues dismiss the association of such poetry with modern agriculture, but as I argue in these chapters, these may be entry points to a greater understanding of agricultural sustainability. Van Gogh's sentiment might seem to tend towards pantheism, yet it conveys his intended contrast with the Christianity of his experience. And since that time, the influence of Christianity has declined in Western societies almost in step with the ascendance of a modern obsession with the environment and in particular, sustainability.[12] So with the irony that often accompanies desperation, the church now advocates sustainability as a newfound virtue.

We can see the church scrambling for relevance when it embraces sustainability as a moral approach to interactions with nature. At the same time, selective acceptance of Buddhist teachings has allowed it too to be portrayed as 'green'. Is either of these associations true? Can they assist an understanding of sustainability? The answer to the first question is probably *no, neither Buddhism nor Christianity has an explicit environmental message*. I consider this statement further and from multiple perspectives in the ensuing chapters. However, notwithstanding that tentative answer, the reply to the second question is a definite *yes, consideration of both traditions can assist understanding of sustainability*. For in the underlying spirituality of these religions we find both the

[12] Timmerman (1992)

source of our yearning for sustainability and the means to practise such forms of it as sustainable agriculture.

Sustainability, like Christianity or Buddhism, means different things to different people. So, rather than walk the well-worn route of seeking a definition of what sustainability really is or should be, the following chapters take a less travelled path to ask whether sustainability could simply be an unacknowledged proxy for beliefs in immortality or rebirth. Such beliefs seem to be among our oldest and, while intended as metaphors for self-transformation, have often been interpreted literally as unchallenged dogma.

Another scriptural misinterpretation may well be the apocalyptic outcomes of our excesses. Sustainability is often a reaction to such fears of environmental catastrophe expressed in terms that mimic Christian eschatology. Predilection to such thought is so powerful that we might ask whether we projected our apocalyptic predilections onto Buddhism as we interpret it in the West. Consideration of such questions also informs the ensuing chapters.

But even centralized and institutionalized religion is dynamic, and the environmental message of the Christian church has shifted from one of condoning exploitation to one of stewardship, equitable land distribution and simplicity of lifestyles. Such sentiments seem to be the stuff of sustainability and can be linked to teachings of universal divine imminence by suggesting that unethical farming practices, for example, are not in keeping with God's law. In Buddhist terms this might simply be seen as ignorance of the complex interdependence of all things, expressed as not acting in accordance with natural law. This essential and recurring theme allows us to reconsider the intent of metaphor and allegory relating to agriculture and nature.

So we are led to ask, is the pursuit of sustainable agriculture a response to mythological and allegorical means of conveying a spiritual message of oneness with all things?

Or, is this interest in sustainability just a new form of pantheism, which sees the divine in all nature and encourages a maternal and interrelated conception of life? All of these mental constructs are valuable means of conceiving a wider reality than that which guides our everyday life. Such considerations lead many to suggest that there is a fundamental conflict between the single-mindedness of mechanistic profit-oriented agriculture and such matters as the ethical treatment of animals as a component of sustainable agriculture.

Rather than divide ourselves into two opposing camps, one for intensive profit-driven agribusiness and the other for ecological sustainability, it seems wise to acknowledge that each has value. And that must surely be superior to the naïve ecological advocacy that has produced impractical theories of an ecologically sustainable agriculture that does not impact on the natural state. Agriculture changes natural environments in all cases and its appropriate practice for the foreseeable future is within that new agricultural-ecology. But it is more productive to approach such matters through philosophy and applied ethics rather than bigoted lobby positions or even religion in the West.

There are also some other questions that we may do well to keep in mind. When we do consider such matters, are we forgetting the assumption of continued growth and its conflict with our yearning for sustainability? Are we seeking to sustain our own lifestyles ahead of others – and isn't this what all organisms do anyway? Rhetorical questions perhaps, but it is from this compromised and confused position that these new ideals of sustainability are being transferred to less-developed countries. In so doing, their experiences and traditions that could otherwise inform our broader understanding of sustainability are unwittingly negated. One such lesson appears to be that the spiritual dimension has been integral to sustaining small-scale agriculture. Such a lesson could work in concert with an intellectual dimension to integrate all of our insights into holistic agricultural practices.

Buddhism is popularly said to offer a means of understanding sustainability that accords with scientific discourse because both share a common base in cause and effect. But is this likely to be true if science is integrally linked to the economic system with its requirements of growth and protection? Even a cursory reading of the Buddhist tenets suggests that to live in a sustainable manner means to meet essential needs and to then apply oneself to developing wisdom. Other teachings invoked in support of environmental messages include the pre-Buddhist ethic of non-violence as a basis for understanding the integrated worldview that is critical to sustainability. Notwithstanding obvious good intentions, we might well consider whether environmental messages are a core of Buddhist insights about cause and effect as a durable natural process amidst the impermanent nature of all things, or whether they are simply examples of ethical guidelines. Once again, conducting the discussion across a wider perspective aids understanding of sustainability.

Without pre-empting the deliberations of subsequent chapters, we may understand the argument by considering that sustainable agriculture is a natural and non self-interested action. In that case it is that latter description - an ethical guideline - that indicates the source of unsustainable actions, where multiple and often self-interested objectives of science and commerce are given precedence. Those who seek to resolve such apparent conflicts by recourse to scriptures appear to either limit their understanding to literal interpretations, or to be disappointed. Is this because the scriptures are oriented to communicating a central message of self-transcendence and simply employ environmental metaphors and allegories to convey that message? I am convinced that this is the case, and that this explains the confusion of well-meaning scriptural literalists who seek prescriptive approaches to sustainability and self-sufficiency.

If we allow that the modern Western environmental stance may be projected onto such exotic insights as those of Buddhism, we can react by seeking some purist version of religious traditions. But if we recall their essential similarity with the insights of Christianity, we may understand why sustainability is an unattainable grail whenever we seek it for our own gain.

Such thoughts are a stumbling-block to some and a folly to others. For me, the almost religious zeal with which even educated participants engage with environmental matters is sufficient justification to take this broader approach to sustainability. However, the justification may turn out to be the insight that develops from this very approach – for our negligence of spiritual development has separated us from the experience of past generations. In the case of sustainable agriculture, perhaps by seeking to sustain productivity we are ignoring the inevitability of change or even working against the espoused objective.

With the uneasy feeling that these questions raise, we can discuss each aspect in more detail and reach the conclusion that in the search for a sustainable agriculture we must ever restless be, until we find our rest in the natural flow of all things. The first of the obvious religious aspects of sustainability is immortality and rebirth, which is considered in the following chapter.

Chapter 2

Immortality:
Sustaining Ourselves?

Two ideas are psychologically deep-rooted in man: self-protection and self-preservation. For self-protection man has created God, on whom he depends for his own protection, safety and security, just as a child depends on a parent. For self-preservation man has conceived the idea of an immortal Soul or Atman, which will live eternally. In his ignorance, weakness, fear, and desire, man needs these two things to console himself hence he clings to them deeply and fanatically.

Walpola Sri Rahula

Is our quest for a sustainable agriculture simply an expression of our deepest fear of our own mortality? Biological science offers little to such a question. Likewise, the religions of the world, which while maintaining platforms on the subject of immortality, quickly erect the shroud of 'belief' before deep inquiry. It has been argued[13] that religious images of immortality can actually be harmful and illusory if not subordinated to integrated approaches to life in harmony with society and nature. However, as with so much of culture-bound analysis, this everyday world not only excludes the next – which seems appropriate, but also excludes the non-Judeo-Christian world. And exclusion from Eastern thought unnecessarily cuts the West off from a most useful means of understanding its own culture. Nevertheless, writings around

---

[13] Dewey (1959)

the subject of immortality in Western culture provide a convenient starting point to answer the opening question.

The weight of words on immortality is overwhelmingly in favour of some form of our continuity, though the details vary and are vague. Yet it is those who argue the non-existence of immortality, and especially those who argue that teachings about immortality are misinterpretations of deep insights, who have more to offer our examination of sustainability. An undogmatic consideration of the subject allows its examination from psychological, historical and various other perspectives that are often denied in authorized exegeses.

Ancient teachings testify to the attraction of the idea. Socrates was apparently interested in the possibility of immortality, Job asked *if a man die, shall he live again?*,[14] and Paul linked the efficacy of faith to the resurrection of Jesus.[15] Indeed, the genius of Christianity is represented in its inclusion of the ancient Mediterranean teachings of immortality as a means of enhancing its early popularity, which incidentally produced such later aberrations as intercession for souls in purgatory. As becomes clearer in later chapters, the allegorical intent of descriptions of eternity and heaven as higher earthly states is largely lost from most popular religion, and this in itself further confounds attitudes to sustainability. But first we might consider where this concept of immortality sprang from.

Belief in an afterlife long predates monotheistic conceptions, and may well be a natural outcome of primitive religions and simple reasoning. Concepts of God are not essential to such beliefs; the idea of God seems to have arisen with increased sophistication within a culture and to then have required instruction or at least explanation.[16] Pre-theistic

---

[14] Job 14:14

[15] 1 Corinthians 15:14-19

[16] Lamont (1959) Page 9

beliefs of immortality might therefore be important psychological supports to life in general. They served as and remain coping mechanisms for consciousness of our impermanence. But afterlives were not uniformly heaven-like – to the ancient Greeks and Hebrews, the afterlife was an unhappy and vague existence that reads a little like Oriental concepts of an unfortunate rebirth.

The early Church allowed emphasis on a desirable afterlife after its prediction of Christ's second coming became untenable. Doctrine was later expanded to allow influence on the afterlife through indulgences. With the Reformation, Protestantism removed purgatory from a part of the religion's doctrines, yet retained an emphasis on eternal reward or damnation according to one's temporal behaviour. In this way Protestantism perpetuated literal misinterpretations of the metaphor of *in earth as it is in heaven*[17] and the role of ethical behaviour in developing wisdom. And it is on earth that the popular focus rests in the current era, taking the form of sustaining environments with a zeal that seems to increase in parallel the gradual demise of societal beliefs in immortality.[18]

Social benefits clearly accrue to a general belief in an afterlife in such times as wars and epidemics, and our behaviour in this world is clearly more easily controlled when we fear punishment in the next. Within such a belief structure, one would appear naïve to act virtuously without hopes of future reward and thus beliefs provide social cohesion that is reliant on codified religious laws. Of course we can act virtuously without such a belief structure or fear of social exclusion. Is this a possible explanation for apparently altruistic actions in seeking to sustain agriculture and other everyday practices while disavowing immortality?[19] It may well be – although as we shall see, such actions may be less

---

[17] Matthew 6:10
[18] Baillie (1933) Pages 36-38
[19] Kirsopp (1922) Pages 21-23

than altruistic. In any case, 'belief' in sustainability has created a burden of clever argument for religious bodies as they seek to accommodate shifts in public attitudes to nature by reassessing their teachings.

The turning point in Christian interpretations of the human relationship to nature is indicated in considerations of the integrity of creation. And of course, integrity is shattered by separate consideration of its components in isolation from each other. The link between this disintegrated worldview and repetitive acts of domination provides one means of understanding the Western culture. In the same manner, conception of the soul as separate from, and superior to, the body fail to describe an integrated person. This unintegrated view produces an unbalanced emphasis on immortality in the face of observed transience, thereby adding to the psychological stress of modern life. Perpetuation of the unintegrated view also leads us into life-extending technologies in modern medicine to ward off the reality of personal impermanence - our own un-sustainability. A telling corollary of this argument concerns modern interest in cloning and in-vitro technologies as a means of overcoming fears of impermanence in the form of 'fatal genes' that arise from the natural variations of sexual reproduction.[20] Used as a means of self-distraction from pursuing the real meaning of eternal life, modern faith in such clonal immortality makes for amusing musings on literal interpretations of such scriptures as *[those resurrected will] neither marry nor be given in marriage.*[21]

Those who believe in a personal immortality, which Kant interprets as securing desired happiness from an invented God,[22] have prosaically described it as more satisfactory than the alternatives.[23] In this pre-clonal yet post-

---

[20] Moltmann (1985) Page 247
[21] Luke 20:35
[22] Kant (1993)
[23] Bliss (1926) Page 56

theistic world, immortality still retains its ancient guises of a continuing personality, name and memory, thereby appearing to confirm Freud's view that, despite rational conclusions about life and death in general, *in the unconscious every one of us is convinced of his own immortality*.[24] Whether we unconsciously see ourselves as immortal or not, we continue to fear death.

The fear of death is apparently endemic in all societies, and in the West in particular death is hidden behind closed doors and euphemisms. The wish to transcend death, described by Freud as *the oldest, strongest and most insistent wish of mankind* has similarly been acknowledged by the perhaps more religious Jung who observed that *rebirth is found at all times and in all places*. Rebirth and immortality may be seen as different beliefs by purists, yet they seem to tap the same psychological aspiration, and both offer us a means of considering the cultural impact of different religious teachings. While a culture was contained through uniform instruction, a continuum between life and afterlife – regardless of its location – codified this deep yearning in various religious forms, with a very high degree of success. Social rules, religious beliefs and education represented a unified system that once defined all of life for most people. But alternative beliefs and modes of education have now extended across the masses and belief in immortality and rebirth has been sublimated at a psychological cost.[25] To take a basic example within our theme, fear of an unsustained food supply may be simply another expression of such anxiety.

Anxiety itself can be seen as a useful mechanism for survival when it functions effectively. It may be effective as an ameliorating belief in an afterlife when it operates as a feedback-loop that allows us to maintain function while always threatened by death. This may be similar to the

---

[24] Freud (1924) Pages 305, 313

[25] Reanney (1991) Page 4

benefits of survival rewards for working in teams, or on long-term projects, or even acting altruistically. Such thoughts must lead modern readers to consider quests for sustainability to be an expression of genetic self-perpetuation in the manner of Dawkins'[26] 'selfish gene' hypothesis. In this way, a belief in immortality or in the virtues of sustainability could be the means by which genetic continuity is made socially desirable. When we recall that the decline in formalized religion in the West is coincident with the rise in interest in sustainability we may observe that irrational processes underpin our unchallenged orientation to this new virtue.

Have we made a virtue out of a vice? The ancient invention of an afterlife might simply be a means of coping with the paradox that immortality is only possible through partial replication of one's genetic material.[27] Such a realization would not only explain historical behaviour, but also should encourage care of all conditions suited to the success of one's genetic successors. I find it no coincidence that this is one definition of sustainability. But when the process is foreshortened and encoded as a denial of death, other psychological conflicts erupt when we inevitably are faced with death. So we cling to either a belief in an afterlife or the virtue of sustainability in order to cope with worldly existence. By the same argument we can suggest that evolution has equipped us with emotions such as love, care and concern for progeny in order to sustain our genes. Such reasoning is evident in supposedly altruistic aims to ensure a viable lifestyle for future generations, sometimes unashamedly expressed in sustainable development rhetoric as *ensuring the future for our grandchildren*.

Securing our grandchildren's survival as a definition of sustainability implies a strong emphasis on a separate self,

[26] Dawkins (1990)
[27] Reanney (1991) Page 10

which itself has arisen from the time perspective of past, present and future. This separation is an analogue of the external world, not the external world itself. Obviously, progressive or linear time allows practical communication within the cycles of natural time shaped by seasons, and by birth and death. While its utility as a mental ordering principle for interaction with the external world is obvious, our acceptance of linear time as if it is a reality in itself limits our perspective.

Time is just a means of describing events in either rhythmic terms as for seasons or linearly for ordering of experience and received knowledge. It is a convention, and while it is facile to describe it here as cyclical, it does have a repetitive nature insofar as certain events such as discomfort from unwise actions seem to recur. Logic tells us that each event is affected by earlier ones and that the exact nature of each event is, like subatomic particles, a probabilistic function because we cannot predict the effects of earlier events. This obviously includes those we call 'sustainable'. The subtle yet significant shift in human conceptions about time probably began in the upper levels of city states that were insulated from the exigency of everyday life, from where it eventually pervaded all social strata. From this viewpoint it would seem that if our view of time is artificial, then so must be our views of sustainability, including such ego-based actions as setting up conditions that favour survival of our genetic material.

Our false view of time is usually traced to the development of writing. While the Sumerian and Egyptian priests documented variations between years, the Druids who used Stonehenge are said to have relied on a verbal tradition with a less specific base for comparisons across time. The Druids, so it is argued, were less captured by time than the cultures that came to dominate the world. Current interpretations are wont to classify the dominant as the aggressive, materialistic, patriarchal religions and the dominated as non-materialistic caring matriarchal religions. Indicators of a unity preceding such dominance is inferred from such Latin word roots as 'material' and 'matter' sharing a common origin with mother (*mater*). Regardless of etymology, this is a useful metaphor for Western culture's domineering attitudes to nature, including the arrogant assumption that we can sustain, that is control, nature.

The shift from natural cyclical rhythms to being captured by a concept of time also supported rebirth beliefs, which were (mis)interpreted from allegorical writings with spiritual intents. For example, the Judeo-Christian myth of creation in which *the spirit of God moved upon the face of the waters*[28] uses the experience of birth as expulsion from the dark amniotic security into blinding light as an allegory for the psychological growth known as enlightenment. This original heavenly abode is also portrayed as an Arcadian Garden of Eden to which we may return and enjoy that pre-Fall warmth and security. We seem predisposed to the mythical imagery of cosmic creation and descriptions of unity. Diurnal rhythms of dark and light are easily assimilated into myths based on birth, death and rebirth - experienced daily in our rising from sleep. The idea is simply extended across a lifetime to become reincarnation. In this way a cyclical understanding is retained, which Jung described as *among the primordial affirmations of mankind.*[29]

Dichotomies such as dark and light typify the dualistic reasoning that defines our rational processes. This characterizes the opposites of the organized religions such as heaven and hell as well as ying and yang, and now, sustainable and unsustainable. It explains the angst at the impossibility of reunifying opposites in rational approaches, as in Blake's 'spectres' and 'emanations' which arose from united

---

[28] Genesis 1:2

[29] Reanney (1991) Page 105

man.[30] Relying on this dualistic approach, utility-oriented interpretations of science further eroded cyclical understanding of time through its popular presentation of the linear process of evolution. The linear approach is clearly useful for this purpose, but it is part of a natural event just as we are. Taken literally, such an understanding of science can be portrayed as a 'second Fall', for it reinforced our separation from the rhythms of nature and ancient mythological explanations. And in so doing it exposed the ego-self to its own mortality. In this way, the feedback-loop of a belief in an afterlife was destroyed by the modern rational process called science, which in turn is now oriented to sustainability research.

If all this were true, desires for sustainability would be accompanied by fears of death or at least denial and removal of reminders of death. If it makes sense that the ego-self by its very creation brings into existence a fear of death, then it would follow that the fear cannot be expunged except by death of the ego itself. So, to function normally, our fear of death must be addressed, either through transcending the ego or by denial. Transcendence of death requires further explanation in a later chapter. Denial of death is simply a superficial solution that conduces to neuroticism, the modern Western condition. Isolation of dying persons, utilization of tens of euphemisms for the word 'death', and even refusal to advise persons of their own imminent death, all assist us to deny death. Thus sustainability may also be conceived simply as a denial of death.

As our evolutionary background has equipped us with an innate appreciation of the patterns in nature in the manner of evocative art, then one part of the overall pattern of all life is, *ipso facto*, death. And as science aims to identify patterns as its means of understanding nature and ourselves, then death –

---

[30] Berger (1914) Page 110

the antithesis of common conceptions of sustainability – is fundamental to scientific enquiry. However, what we have come to refer to as science, now takes a mechanistic approach and can ignore the inevitability of death of all things as it seeks to manipulate part of nature to suit our preferences. In such a way, science becomes the unwitting servant of delusional permanence.

Where a broader non-manipulative definition of science is taken, an understanding arises that is part of the *causal dependence* insight at the heart of Buddhism, as is expanded later. This is possibly what Jung meant by 'synchronicity'. Such a broad scientific view reveals the complexity of all natural systems while recognizing their interdependence with all other things, even down to the level of thoughts. One outcome of this conception is that human consciousness itself could be a feedback mechanism that contributes to maintenance of the overall system – and that would be a means of understanding our concern with sustainability. From this perspective, sustainability is a natural action of humans rather than one of self-protection.

Sustainability seems a natural response to environmental awareness. But the usual narrow definition of sustainability as sustaining our comfortable state cannot be called natural – it is, if you like, an unsustainable approach to sustainability. It is based on selfish intentions for more and more at the expense of other components of nature. One indication of our efforts to deny the illogicality of such 'sustainability' is our modern comfort being based on hiding death so far as possible. However, such comfort is short-lived as *the happiness that comes from an achieved equilibrium situation lives only in the 'now'; and it is the unhappiness of another stable, non-equilibrium state that thrusts [our] awareness into time.*[31] We are thrust back to reality by death and other unsustainable

[31] Reanney (1991) Page 233

situations as part of the feedback mechanism of the overall system. If we learned the cause of such un-sustainability, we could understand more about sustainability.

From this discussion, we might conclude that our desire for immortality is one possible aspect of our infatuation with sustainability, as this chapter has indicated through its three main points:

- The various cultural forms of beliefs in immortality or rebirth share two defining characteristics; they misinterpret allegorical representations of eternity and heaven, and they have provided a coping mechanism for everyday life lived in the face of death. Institutional promotion of an afterlife has disaggregated understanding of life, making it appear to be under constant threat of being unsustainable.
- The rise of sustainability as a virtue has shadowed the decline of religious influence in the West, causing scriptural revisionists to embrace secular values through a redefining of nature-dominating behaviours while maintaining arrogant attitudes that assume we can manipulate nature to suit our own ends.
- Broad scientific enquiry into the processes of nature recognizes cycles of life and death, but narrowly conceived mechanistic science is easily oriented to forestalling change, and death. While we can conceive human consciousness as a cosmic feedback mechanism that mollifies excessive intervention in natural processes, it is difficult to see sustainability arising from selfishness.

So our search for sustainability may partly be a quest for immortality - and in that context, futile. But it could also imply that a reinterpretation of religious heritage is occurring in the light of science and other insights, and this may well represent a revolution in thinking. So, it is to this subject of changing religious influences that the next chapter is addressed.

Chapter 3

Agricultural Theology: Why we are Fascinated with Sustainability

People like us, people who believe in physics, know that the distinction between past, present and future is only a stubbornly persistent illusion. … No one is able to achieve this [release from pain] completely, but the striving for such achievement is in itself a part of the liberation and a foundation for inner security.

Einstein

While sustainability may well be a surrogate for immortality, we may also see it as a biostatic feedback mechanism that mitigates our earlier environmental excesses. This feedback mechanism is fuelled, in Judeo-Christian terms, by guilt and self-interest. Guilt, and even self-interest, can be powerful motivators for compassionate action, but as I argue here, logic is more useful. And insight is better than logic. While recognizing the inherent difficulty of separating insight from logic, this chapter makes an attempt to apply deeper Christian experiences to past errors in our relationship with nature. It also clarifies the intent of some of these misunderstood Christian doctrines.

In analysing the theology of human relations with nature, the Genesis creation myths are for me as for many, a primary source. Here I follow the detailed work of Moltmann[32] who without actually saying it, contextualised Christianity in the modern secular ecological apocalyptic view.

---

[32] Moltmann. (1985) Page xi

He stops short of comparing Christianity with the new form of salvific nature worship marketed as 'environmentalism' or its synonyms, but is cognizant that the salvation offered by 'ecology' (Greek: 'doctrine of the house') could be a metaphor for God the Creator indwelling in his creation. This immanence is a central element of Moltmann's worldview in keeping with orthodox Judeo-Christianity.

The alignment of theological and anthropological perspectives in ecology may be seen as a means of bringing us back to our home in nature. As agriculture is the major land-disrupting contact we have with nature, the alignment reveals the collective subconscious that underlies sustainable agriculture. This revelation is valid even if a Western person claims not to be religious, or subscribes to the naïve claim of being unaffected by the Judeo-Christian basis of those societies.

In the beginning, so Western culture taught, a cosmic hierarchy led by God as Creator posits we humans as independent clones of God. We are therefore superior to other elements in nature. This conventional view in Christianity is now being challenged as are other diverse relationships of God with creation, including the united psychological traits of the Trinity. But such a conventional theology requires what I have elsewhere referred to as 'experiential knowledge' (*sapientia*)[33] or 'participating knowledge'.[34] 'Experiential knowledge' here means the knowledge developed from internal mental development, sometimes referred to as mystical insight, and differs from the dualistic subject-object reasoning of rational thought. Such insight involves understanding an object as integral with surroundings, rather than as the sum of its individually analysed components. Remembering our participation in community with nature is

---

[33] Falvey (2004) Chapter 9

[34] Moltmann (1985) Page 1-19

difficult when we attempt to be impartial scientific observers of nature and its interactions.

This community, or communion if you like, with and within nature is the indwelling of God in creation expressed in such terms as *on earth as it is in heaven.*[35] Christians would understand all beings as finding bliss in nearness to God, who is in all things. The concept of immanence, which pervades all religions, leads to the fundamental point of spiritual teachings: becoming closer to God. A creative spirit of God, the 'giver of life', that is the Holy Spirit, can be understood as God's manifestation in all living forms. Thus theological and ecological doctrines coincide.

If, on the other hand, God is not seen in the world and the world in God, then *nature is stripped of her divinity, politics becomes profane, history is divested of fate.*[36] Such is not the message of the Old Testament. It negates the central message of a God leading his creation towards him by means of the Holy Spirit. Whether or not we accept the conceptions of 'God' and 'the Holy Spirit', analysis of human thoughts and actions points to the same conclusion, that the spiritual essence we all seek is within each of us. Realization of that essence brings insight into the interconnectedness of all things –similar to, yet more far reaching than, that of ecological advocates.

Unifying disparate environmental concerns into an ecological crisis in industrialized Christian countries has been a triumph of capturing public awareness. This has been made possible by our inherent fear of change, death or un-sustainability as described in Chapter 2. If the 'ecological crisis' is viewed for the sake of argument as a modern myth to replace judgement day or the apocalypse, we may see our psychological predisposition to it. This makes the traditional Christian assumption that we have a right to dominate nature

[35] Luke 11:2
[36] Moltmann (1985) Page 13

more and more politically incorrect. It would follow that any of our work to further understand nature in order to engage in new ecologically destructive actions is a form of destroying the divine, with its attendant anxieties.

We might consider then that redressing of ecologically destructive actions requires social and spiritual change rather than technological solutions. Science is now impugned, not in itself but through its application to an economic model that assumes continued growth in consumption. Pre-Renaissance assumptions that equilibrium as distinct from growth was to be maintained, offer us a glimpse of a societal approach to sustainability. But this changed with the scientific insights that we glibly paraphrase in Francis Bacon's words, *knowledge is power,* in which knowledge now equates with 'progress'. 'Progress' was defined in hopeful terms throughout the nineteenth century. Then it was gradually assumed that the absence of progress, particularly in economic terms, was at a kind of death. In the same way, growth became the vitality of life and so sustainable growth became a modern interpretation of progress. It is this worldview that encourages us to view ancient civilizations as 'under-developed' or 'developing', thereby revealing *the mindless imperialism of this ideology of progress, which judges everything on the basis of its own condition and which aims merely at its own hegemony.*[37]

By stepping back from assumptions of growth, we are able to re-view such anomalies as the peaceful yet domineering mandates of Genesis. Where man is said to be made in God's image with dominion over all in the earth,[38] we may choose to see a peaceful and responsible ruler, rather than one of uncaring and rightful exploitation of the earth, plants and animals. This interpretation accords with the attendant blessing that man should enjoy and live from the fruits of the

---

[37] Moltmann (1985) Page 28

[38] Genesis 1:26

earth. The subsequent reference to tilling and keeping – *to dress and keep [the Garden]*[39] – may be interpreted as sustainable agriculture.

The West's cultural heritage has primarily equipped it to conceive such matters theistically. The Genesis story may be interpreted as advocating respectful administration of an asset entrusted to humans by God. Man serves his ends rather than our own. The final act in the creation myth, God's Sabbath, indicates divine pleasure. So it can be concluded that *even without human beings, the heavens declare the glory of God.*[40] Thus nature is *'very good'*,[41] not man alone, who is but one aspect of nature. Christianity has inherited this Judaic insight in which humans are one with all creation as it anxiously awaits the Creator's will.[42] Western religious heritage has favoured the alternative domineering interpretation from the same scriptures and has thereby subverted an important aid to personal insight to a materialistic intent – including attitudes to sustainability!

The above wider interpretation of Christian cosmology, which extends a transcendent Creator into a presence within each person, has been supported by the liberation of thought by science. However, the church also seeks its own sustainability in material form and is consequently at odds with the original meaning of immanence – personal spiritual development. As so much of the West's intellectual resources have, in recent centuries, been filtered through science and its philosophy, we might expect that a unification of religion and science would produce a new revelation. However, the Protestant demarcation of truths about creation have hindered religion's reunification with science – until now when *scientists are also slowly beginning to discover that Christian theology is not*

[39] Genesis 2:15
[40] Moltmann (1985) Page 31
[41] Genesis 1:31
[42] Romans 8:19-21

conserving antiquated world views, but that it is a partner that deserves to be taken seriously, both in the sphere of cosmology and in the realm of social practice.[43]

Simultaneous interrelationships are more complex than we can conceive in purely scientific models. Integrative thinking can perhaps accommodate both theological and scientific concepts. Such intellectual broadening is more caring of all nature and necessarily includes sustainable agriculture. A theological-cum-scientific approach transcends political outcomes, for environmental excesses are the hallmarks of all materialistic cultures, be they Marxist or Capitalist. The first step in this dual conception of our role with respect to nature is a psychological adjustment. This is uncomfortable, as a new fusion of the Christian ethic with science is often viewed with hostility and as a threat.

Living as 'part of nature' is more than what is popularly called 'nature-loving'. It includes complex theological thought that offers one means of correcting a false view of reality. Rather than controlling nature to produce food, for example, living as part of nature implies that actions be conducted within natural cycles. In practical terms this means producing food with minimal interference with natural rhythms, as many environmentalists say. But living as part of nature could also be interpreted as understanding nature in more detail than the crude perceptions our human senses allow. This squarely opens the way to the application of science to sustainable agriculture. The alternative is to act arrogantly towards the nature of which we are part. According to both the theological and ecological worldviews, such arrogance may yet result in an apocalyptical outcome.

The above argument suggests that an impending ecological crisis can only be averted through a more sensitive relationship with nature. Basic human rights to food and

---

[43] Moltmann (1985) Page 34

habitation would fit within the relationship. In scientific terms, awareness of natural processes might be emphasized. In the same manner that patients can effect their own healing in psychosomatic medicine, our search for spiritual reconnection to natural rhythms may avert the apocalypse. By considering ourselves a part of nature, we can overcome a 'having' approach to things (including even our own bodies), and can assume a 'being' nature. This allows rediscovery of traditionally suppressed aspects of a whole life.[44] While these conclusions arise from a Western perspective, they are remarkably congruent (as presented in later chapters) with those that arise from Eastern perspectives, including those that do not rely on God as part of their explanations.

Working within the Christian paradigm at this time, I would like to examine what the concept of God contributes to our knowledge of nature – rather than what nature contributes to knowledge of God as is usually done. From the conventional perspective, the world is conceived as divine through revelation by God, and that revelation itself renders God universal. From the alternative viewpoint, the early Christian natural theology that ignored extant Stoic interpretations of natural essences produced a contingent view of reality. God was only partially knowable through nature, because scripture and faith were also necessary to achieve the *perfect fellowship with God*[45] in the eschatological view of a promised kingdom. Thus Christianity came to view nature as containing evidence of God rather than as God's total revelation. We may reinterpret the whole Bible from this reconstructed 'nature' perspective, as in the continuing Earth Bible project, which also includes elements of the New Testament.[46]

---

[44] Moltmann (1985) Page 52

[45] Moltmann (1985) Page 57

[46] Habel and Balabansky (2001)

If we consider only the New Testament, as some Christians are wont to do, we find that it accepts the Old Testament metaphor of the world as God's creation.[47] The Holy Spirit is the link between the elements of creation and the Creator. The Holy Spirit also transcends intellectual understanding by drawing us into communion with creation. A return to these essential teachings may yet be possible for *the monastic traditions of the Orthodox church and the Hasidic traditions of Judaism have preserved these splendid concepts. Today they must be rediscovered and translated into the practical dealings of human beings with created nature. They are all suited to overcome the one-sided and impoverished attitudes of people living in the modern industrial world.*[48] If it serves creative didactic purposes, God's creation of the world out of nothing may yet be a useful conceptual device for many in the West.

Creatio ex nihilo, usually conceived as an emotionless act, may also be interpreted as an act of love by distinguishing sacred 'creation' from profane 'making', and 'work' from 'image', in the language of Genesis. New Testament projections of God as love require God to be both supreme substance and supreme subject as conceived in the Trinity. The sum of creatures is not the same as the Creator, as pantheism might suggest. Rather, the ultimate objective of all creation is to find God. Apocalyptic reconciliation is therefore with God in a *new heaven and new earth,*[49] here and now.[50]

The role of Jesus as reconciler confirms this metaphor. If Christ is one with God and saviour of creation, then we may argue that he is also the source of all creation. Salvation then is an awareness of God's sustaining the world through Christ.[51] The power of the Holy Spirit in creation includes the Hebrew

---

[47] *eg* Matthew 19:4 and Romans 1:25
[48] Moltmann (1985) Page 71
[49] Isaiah 65:17, Revelations 21:1
[50] Moltmann (1985) Page 93
[51] Hebrews 1:2

'Shekinah' (that which dwells). It is also the indwelling of God in our own bodies, *the temple of the Holy Spirit.*[52]

The Trinitarian nature of God may also be expressed as, *the Father is the creating origin of creation, the Son its shaping origin, and the Spirit its life-giving origin.*[53] Separation of these elements has brought about a modern dilemma; emphasis on transcendence produced the deism of Newton, while emphasis on immanence produced the pantheism of Spinoza. The former has proven more socially powerful, and the masculine or lordship role of God has dominated the feminine or world-soul aspect, so that nature is conceived as spiritless by the mechanistic analytical sciences. When complex systems are assumed to be a compilation of segregated systems or as chaotic patterns, as is the case for much of our mathematical approaches, we omit this essence. This applies whether we see God as a metaphorical tool for our self-discovery or as a being in his own right.

Our initial insights along Christian lines can be expressed as *the first-fruits of the Spirit,*[54] and as the anxiety of all creation which includes us in longing.[55] These thoughts point to a solidarity with nature within the reality of transience. They include an active developmental aspect, a move toward the transcending of everyday changes. In contrast, pantheism produces ambivalence, for everything appears of equal importance when God is complete in all things, and when there is neither differentiation nor progression. So we may conclude that pantheistic and mechanistic approaches each provide only a partial picture of a system perpetually subject to the Christian God. In our example of agriculture, our meagre understanding of its countless interacting processes make a nonsense of attempts to

---

[52] I Corinthians 6:13-20
[53] Moltmann (1985) Page 98
[54] Romans 8:23
[55] Romans 8:19-22

sustain anything by our management, unless part of an overriding imperative to enhance our personal relationship with God in this theistic conception.

However, such a 'knowing of God' is fractured, as introduced above, by our adherence to an artificial concept of linear time. We are blinded to our repetitive and unwise actions in temporal terms that are sometimes described as cycles, and to the non-existence of time in the transcendent sense. The Greek source of insight about time is indicated by a fragment of Parmenides'[56] writings in which 'being' is considered both divine and eternal such that 'becoming' or 'non-being' cannot exist. Rather than time passing, events are seen to pass the eternally present. If the secular world is conceived in terms of the linear passing of time then it becomes separated from the natural world and its rhythms. This artificial separation is the paradigm in which we commonly seek to sustain agriculture, and it is a futile endeavour. It ignores cycles of which we are part and of which we may be agents. It essentially views nature as either static or manipulable according to human whims. If we could understand sustainable agriculture in terms of a transient event passing through time, then time could be *brought into harmony with the laws of life and rhythms of nature*.[57] Sustainable actions would be those which are coincident with these rhythms.

As with our mechanistic view of time, so are our views of space. Our once finite world has been expanded by science to a universe so vast that it is beyond our comprehension. This has fuelled our nihilistic tendencies, for it has accelerated the loss of holy areas and of sacred space.[58] Civilisation developed from religiously-conceived agricultural enclosures and as a sacred sense of place. But as these came to be defined in

---

[56] Cornford (1939) Page 30
[57] Moltmann (1985) Page 138
[58] Moltmann (1985) Page 143

Cartesian terms as utilitarian objects, the concept of the holy 'living space' of humans lost its connotations of interaction with nature - in clear contrast to the poet in Psalm 104 who considered the elements of the environment and their interactions as all components of life, even within the conception of a creator.

The creation account also offers a spatial hierarchy. The heavens have their function,[59] the spaces of sea, air and earth suit the beings which live in them,[60] and the environment modified by these living things provides the world for which we are created, with plants as our food.[61] *If we think in terms of environment and biotopes, the construction of the first creation account is astonishingly clear and logical. Modern reproaches that it is the mere outcome of mythical speculation, or that it displays a naïve knowledge of nature, are quite wide of the mark.*[62] The relativity of space and time is easily forgotten in our communication, and when we insist on their specificity we make erroneous decisions about reality, including sustainability.

The Renaissance revived ancient speculations of the Greek philosophers concerning the nature of space - whether it is an extension of the objects within it (Aristotle) or the receptacle for those objects (Plato). Subsequently, the concept of infinity supported a view that matter is infinitely extended in space, which caused Spinoza to advance the pantheistic view that all matter must be divine. Descartes, with the advance of mathematics, heralded the separation of the concept of God from nature, and limited God to the soul while mathematically objectifying all matter. Since the Renaissance, science has continued to consider nature as separate from the divine. This explains why modern agricultural science has

---

[59] Genesis 1:6-8

[60] Genesis 1:9-12, 20-22

[61] Genesis 1:24-28

[62] Moltmann (1985) Page 149

little value beyond productivity - hence long term productivity is a common definition of its sustainability.

In contrast, true agricultural sustainability is a manifestation of 'heaven', harking back to the earliest concepts of sacred space providing life-giving sustenance. If heaven is part of the created world, yet distinguishable from the visible component, as is expressed in early Christian creeds that refer to *all things visible and invisible*, then God's immanence in the world is heaven. In Trinitarian terms, God the Father dwells in heaven, the Son on earth, and the Spirit in both, bonding the whole creation of heaven and earth. Heaven becomes *the openness to God of the world he has created [where] earth means the reality of the world which is knowable [and] heaven means God's potentiality for the earth, which is unknowable.*[63]

However, when the church changed the original prayer for God's will *on earth as it is in heaven* to a hope for one's soul to go to another place called 'heaven', teachings about salvation of the soul replaced those about God's immanence in the world. By limiting God to a heaven separate from earth, the creative (and therefore also the destructive) potential of God was also separated from everyday existence, and so false views of sustainability arose as logical possibilities within a selfish ethic. These selfish views also provided the basis for apocalyptic teachings of earthly annihilation, tantamount to the rejection of God. Such environmental viewpoints are as un-Christian as the separation of heaven from this life and are *merely the ideological self-justification of nineteenth- and twentieth-century man, in his conquest of the world, his exploitation of nature, and his self-deification?*[64]

The Hellenistic and humanistic summary that 'man is the measure of all things' influenced and influences Christian interpretations. If, on the other hand, we are not the objective

---

[63] Moltmann (1985) Page 165
[64] Moltmann (1985) Pages 193-194

of the cosmos, of evolution, or even of the world, our purpose in the Christian model is found in God, as is the meaning of all things – in Gnostic terms, this means that God and man are one. This is the essence of Christianity, though it is seldom appreciated, and it is entirely consistent with the insights of other traditions when they too are stripped of their accretions. So, if the Christian (and all other forms of) God is immanent in the world, teachings that divine creativity is limited to the beginning and that eschatological teachings only concern redemption, impose a restriction on God's *creatio continua*.[65] God's continuing and continual creation is true 'sustainability'.

However, once again we are confused by our prison of time. Our creation in God's image with power to rule over other beings based on the Genesis myth[66] is used to misinterpret New Testament references to insight rendered into everyday language.[67] Literal readings of the New Testament commonly produce aberrant beliefs, for many passages use the Old Testament as a source for metaphors of the ideal (restored, saved or enlightened) man. The man made in the *image of God*[68] is Jesus, to whom *all authority is given in heaven and on earth*.[69] The New Testament teaches that by following Jesus we are restored to our Adamic condition in the image of God, with the incumbent responsibility to sustain the world. This is reinforced in Luther's words *whereon thy heart is set and whereon it relies, that is in truth thy God*.[70] So it would seem that, if we set our heart on sustained food production or even immortality as considered in Chapter 2, we worship a false God.

---

[65] Moltmann (1985) Page 212
[66] Genesis 1:26-27; 5:1; 9:6
[67] Matthew 28:18
[68] Genesis 1: 26
[69] Matthew 28:18
[70] Moltmann (1985) Page 234

Once again I feel bound to emphasise that the language used in this chapter follows that of Christian cosmology and centres on God. As shown in later chapters, the use of God in the discussion is not essential either for an understanding of sustainability or for an understanding of reality. 'God' infuses the tradition from which Western society sprang, and to reject the message because the language is unpopular today is to throw the baby out with the bathwater. The false God who exists before the creation and who rules the immortal afterlife shatters the integrity of creation. This fracturing produces the artificially separated components that we seek to dominate and sustain. To consider the soul separate from the body is to similarly misunderstand ourselves. This leads to powerlessness and the life extending technologies of modern medicine. Yet even the Old Testament assumed an integrated human in its introductory words, *man became a living soul*.[71]

The separation of body and soul as considered in Platonic views is countered in Paul's proposition to the Greek Church of Corinth with the words, *the Lord [belongs] to the body*.[72] Yet that Platonic view has been advanced by the church, and even until recent times, by science. As integrated beings, interconnected with all other aspects of nature, we embody the creative spirit in imaginative and inventive communities. That spirit has been defined as love, a condition capable of happiness and suffering in a life lived without reserve, one leading to transcendence of everyday vicissitudes into an eternal life.[73]

Such 'love' includes accepting natural changes in the environment rather than chasing 'sustained outputs'. Sustainability is also informed by a consideration of creation being *performed for the sake of the Sabbath*,[74] rather than the

---

[71] Genesis 2:7
[72] I Corinthians 6:13
[73] Moltmann (1985) Page 275
[74] Moltmann (1985) Page 277

Sabbath being God's rest from the work of creation. This feast of rest continues, as no night followed the last day of creation. This implies an acceptance of and even a resting in dynamic nature as it is. The Sabbath may also be viewed as an allegory for periodic resting periods, such as agricultural fallows. Resting 'in' his creation rather than 'from' it, God is a metaphor for that stillness that humans seek. Thus we can understand St Augustine's description of the human heart that shall *ever restless be until it finds its rest in Thee.*

The resting of God's presence is the 'sustaining foundation' that preserves nature from obliteration. The Sabbath is a source and symbol of inner liberation, matching the outer liberation told in the Exodus story. Redemption of creation is symbolized in the Sabbath rest as the feast of creation. When we consider the seventh year fallow, *in which you shall not sow your field or prune your vineyard,*[75] we see that the sabbatical is an appropriate Christian ecological response today in both spiritual and practical terms.[76] The contrast with the attitudes that accompany modern intensive agriculture may explain why sustainable agriculture remains elusive when it is limited to maintaining profit from the traditional capital base. Of course, we may also postulate a world fed sustainably from intensive fermentation products and hydroponic agriculture, but that is not the focus of agricultural sustainability in our current context.

In this chapter we have examined a number of sacred cows, and sought to expose the original intent of the Christian scriptures. These were interpreted in a way to assist our exploration of sustainability. The essence of the arguments may be summarized as follows:

- Christian assumptions of human rights over nature seem to have misinterpreted the intent of the creation myth,

[75] Leviticus 25:1-7
[76] Moltmann (1985) Page 296

which includes emphasis on the interconnectivity of all elements; deeper understanding of this interconnectivity indicates that it is the source of bliss that is otherwise described as nearness to God.

- False conceptions of time and space determine the falsity of some actions labelled as sustainable; these include mechanistic manipulation of components of an impossibly complex system that ignores the divine essence in ourselves, thereby producing the indeterminate angst prevalent in modern life when it seeks sustainability through continued growth.
- Environmental apocalyptic scenarios mimic Christian eschatology; they distract from the central Christian message of the re-unification with God that affirms the integrity of nature, and the metaphorical intent of Jesus' life as being in and of God, which itself is a metaphor for sustainability.

From the perspective presented in this chapter, we can see that the insights of Christianity have suffered gross misinterpretation, to the extent of licensing selfish environmental exploitation. While agricultural sustainability may be one means of redressing the excesses of that license, our false view of reality – according to these Christian insights – continues to limit our effectiveness. This derives directly from our motivations. If we are greedy, overly competitive or just plain deluded, then we mistake sustainability for something that we can control. Control – in the form of security afforded by immutable rules for the minutiae of everyday life – is the preferred approach of those who seek to interpret scripture in a literal manner. Such control is considered in the following chapter.

Chapter 4

Literal and Historical Christianity and Agriculture: Our Manipulations and Our Undoing

We read the complaints of great men in every century about the customs of their age. They always sound as if they referred to our own age, for the race is always the same. At every time and in every art, mannerisms have taken the place of the spirit, which was always the possession of a few individuals, but mannerisms are just the old cast-off garments of the last manifestations of the spirit that existed and was recognized.

Schopenhauer

The Western relationship with the natural environment may well have rested on poor theology since the Renaissance, yet the error could be a means of bringing Western culture to a new insight into human's role within nature. However, Christianity is unlikely to be a source of societal interest in sustainability since sustainability has apparently not penetrated far into everyday Christianity.

This observation itself raises the prospect that Christianity is today following rather than leading secular morality. In so doing it is seeking to modify the relative emphases of its teachings. Disregarding this motivation until later, this chapter discusses the everyday literalistic re-interpretations that are belatedly creating an eco-Christianity, of which Bible-sanctified sustainable agriculture is part. Such re-interpretations might be discounted as self-serving and blinkered misunderstandings of scriptural intent but they have in fact assisted development of the very science that we venerate today.

We may trace the Western advance of science to Luther, though he would have been surprised by the association. When Luther's insisted on the Bible as *sola scriptura* and rejected commentary on its allegory and symbols he licensed literalism, thus he advocated *scripture without any gloss in the sun and the whole light from which all teachers receive their light, and not vice versa.*[77] Calvin followed the trend and the previously accepted interpretations of nature as another revelation of God's work were lost to a group that became associated with rising power. Protestantism suited the political and economic times, and in fact one may read Bacon as reacting against this by accepting both scripture and creation as God's revelation. Galileo too saw God's work written in mathematical language but knew it as human's language not, as Descartes apparently did, as God's formulae. This history of science and religion challenges the chestnut that science produced secularism by encouraging disenchantment with religion. It implies that it was in fact Protestantism that produced disenchantment by stripping away allegory and symbols.[78] By omitting nature as well as allegory from the religion, Old Testament authors were open to overly literal interpretations.

The Old Testament is replete with 'environmental elements' such as, descriptions of paradise, the flood, the destruction of Sodom, the Plagues of Egypt, the parting of the Red Sea, the story of the burning bush, Joshua calling on the sun to delay setting, Jonah and the whale, Joseph and grain storage.[79] These are used in many instances to justify an association of Judaism, Christianity, and of course Islam, with nature. But the real meaning of the stories rests in their

---

[77] Luther (1484)
[78] Harrison (2001)
[79] Gerstenfeld (1998)

didactic purpose of explaining the self-transcending goal at the heart of each religion. The same is true of scriptural references to agriculture, which forms a common store for metaphor and teachings in religion.

Agriculture provides themes and examples for Christian and other teachings, not because agriculture itself is special (though access to food obviously is),[80] but because the teachings were formulated in an era when agriculture was the major occupation of advanced societies. The Bible is replete with agricultural references to justice, compassion and stewardship. Justice is represented as equity of access to the natural resources required for family food production,[81] compassion is represented as sharing with and caring for all nature,[82] and stewardship is represented as respecting and valuing the natural environment through responsible use that considers future generations.[83]

Specific aspects of agriculture, which might be claimed as elements of sustainable agriculture, are embedded in such biblical teachings as those concerning land distribution, agronomic practices, mortgage management, the role of work and technology, and off-farm community responsibilities.[84] An equitable land distribution system is described as including an enforced revision of land ownership every 50 years in order to allow continuity of family farming and adjustment of debt.[85] And fifty years is incidentally the time frame for the generational change in land managers required today to effect significant changes in environmentally damaging practices in agriculture. The tendency to consolidate small family farms into larger farms and to thereby remove families from

---

[80] Falvey (2004) Pages 85-118
[81] Amos 5:24
[82] Psalm 145
[83] Genesis 1:28; 2:15
[84] Banks and Stevens (1997) Pages 401-404
[85] Leviticus 25:23-24

agricultural activities is warned against in the words, *woe to you who add house to house and join field to field till no space is left and you live alone in the land.*[86]

In addition, we can find biblical references to agronomic practices that include soil care and enhancement, the production of healthy agricultural produce, and minimal interference with the natural environment.[87] We find words reflecting modern views of agriculture as fulfilling,[88] as benefiting from technological innovation,[89] and as doing so without selfishness, greed or ignorance. 'Beyond the farm gate', the functionality of communities[90] is based on shared responsibilities and labour, for example in the development of essential buildings and infrastructure. This is in obvious contrast to the individualistic approaches of modern agriculture.[91]

We may extend this revisionist zeal into stewardship, which is a common link between Christian ethics and sustainable agriculture. This can be paraphrased as *stewardship of God's earth for the benefit of our children's children.* Setting aside the implications of genetic immortality included in such statements as discussed in Chapter 2, stewardship may be discussed as a distinctly Protestant, or at least post-Reformation, term that has been reunited to domesticity with its siblings, 'economy' and 'ecology'. These latter share the same Greek derivation. Proponents even trace stewardship back to Joseph working as a steward in the house of Potiphar,[92] and then broaden it to acknowledgement of God as the ultimate owner of everything[93] who entrusts all creation to

---

86 Isaiah 5:8
87 Jeremiah 2:7
88 Genesis 2:15
89 Exodus 35:30-31
90 1 Corinthians 12:12-13
91 Banks and Stevens (1997) Pages 401-
92 Genesis 39:8
93 Psalm 24:1; 50:10

the care of humans. Such an interpretation is said to accord with the discretionary power accredited humans to care for the earth.[94]

We can go further with these interpretations. If individuals' rights are subordinate to those of the community, then personal accumulation of natural resources can be understood as misuse of a divine gift. New Testament teachings on stewardship also extend to sharing personal time, ability and finances for the benefit of others.[95] This is linked to the evangelical objectives[96] of sharing assets[97] and assisting the poor.[98]

Biblical encouragements of simple lifestyles are exemplified through caring for[99] and not ignoring[100] a poor neighbour. Jesus' own lifestyle exemplifies detachment[101] from the anxieties of possessions and in sharing with the poor. This was apparently practiced by the early Christians.[102] Self-control[103] and generosity[104] are fruits of the spirit that exhort everyday vigilance not to live above the lifestyles of one's neighbours[105] and to assist the needy.[106] In terms of not stressing personal gain, such teachings might appear consistent with some conservative advocates of sustainable agriculture springing from deep-ecology.

An example from a practical interpretation of biblical injunctions in nineteenth century agricultural education in the

---

94 Genesis 1:26-29; 2:15
95 Matthew 25:14-30
96 1 Peter 4:10; 1 Corinthians 4:1; Titus 1:7
97 Acts 2:44-45; 4:32-35
98 Acts 11:27ff
99 Psalms 112:1, 3-5, 9; Proverbs 31:20
100 Amos 6:4-7
101 Luke 12:22-34
102 Acts 2:32-35, 44-46
103 Galatians 5:23
104 II Corinthians 8:8-9
105 I Corinthians 10:28
106 II Corinthians 8:1-5

USA illustrates the difficulty of applying personal ethical views and literal interpretations on a large scale. The Illinois Industrial University aimed to produce 'men of Christian culture … able and willing to lend a helping hand in all the great practical enterprises of this most practical age',[107] by which they meant modernizing agriculture. But their approach to agriculture was to reverse the curse of Adam[108] by liberating 'the toiling millions of mankind'[109] who 'eat their bread in the sweat of their brows' as God had decreed when Adam and Eve were cast out of paradise. The tenor of the movement and its hymns, are appealing and reminiscent of the moral tone of modern fundamentalist rhetoric. And it seems to make a positive contribution to practical agricultural education. But it misses the redeeming intent of the Eden story.

The moral tone of the above interpretations is worthy, yet it is but a tiny and biased part of the profound and mind-changing message intended by the scriptures. However, in this chapter we focus more on the literal as we seek the origin of sustainable agriculture in modern Christianity. The argument goes like this: in relying on God to supply all needs,[110] the Christian life is one of material modesty[111] rather than desire for profit,[112] provided the basic necessities of food and clothing are met and any other possessions are viewed as unnecessary though accepted with gratitude.[113] Such interpretations are not new. Sects have often countered avarice with codified forms of renunciation.

One popular set of recommendations for simple everyday Christian life has been developed by the Central Committee of the Mennonite Church. It endorses family

---

107 Smith (2004)
108 Genesis 3:17-19
109 Smith (2004)
110 Philippians 4:11-13
111 1 Peter 3:3-4
112 1 Timothy 3:8
113 1 Timothy 6:8

vegetable gardens, walking or cycling, avoiding fashion, patching clothes, sharing accommodation, and eschewing recreational shopping.[114] Laudable so far as they go, it is easy to forget that such moral restraint is one spiritual exercise among many that aims to effect a fundamental change of mind. Such actions contribute to sustainability to an extent.

The genius of the Bible includes its openness to consideration from varying perspectives, not just the literal, which can appear quite limiting. Other intellectual approaches include the historical or Midrashic styles in which exegetic scriptures were written to accord with prophetic teachings. Midrashic interpretation of significant events or persons can provide deep insights, such as was touched on in Chapter 3. Such analysis relies less on theology than social history, and provides yet another avenue for consideration of sustainable agriculture, to which we now turn.

In contrast to applying modern interpretations to ancient words, we might consider the Old Testament to be an unfolding of human understanding of spirituality. Expressed in Christian terms, this is God's progressive revelation to humans. In this way, the Old Testament can represent a biography of God. This leads to new insights into the intent of scripture. In taking that approach, multiple and conflicting personas are revealed of an often inconsistent and demanding God. This should not surprise us, as the concept of God was and is always being refined. Far from being sacrilegious, this approach allows us to examine references to nature and hence sustainability in terms of the developing self-awareness of consciousness. Within such a context, literal interpretations can also be considered. And whether we acknowledge it or not, modern Western society continues under the influence of *the improbably unexpurgated biblical page [where], God remains as he has been: the original who was the Faith of our Fathers and whose*

---

[114] Banks and Stevens (1997) Pages 401-404

image is living still within us as a difficult but dynamic secular ideal.[115]

It is within such a refreshing approach that we may understand references to agriculture and nature in Genesis, including in the Adam and Eve myth.[116] Adam's punishment for accepting the forbidden fruit was, *cursed be the ground because of you; by toil shall you eat of it all the days of your life: thorns and thistles shall sprout for you. But your food shall be the grasses of the field; by the sweat of your brow shall you get bread to eat, until you return to the ground.* This is simply the reciprocal to the curse on Eve of childbirth pain, human friction and conflict.[117] Realising that this was a period of societal conflict between pastoralists and farmers, we might simply see the words to Adam as a description of the incremental work that agriculture required compared to the earlier hunting-and-gathering or the coincident pastoralism.

The friction between low intensity pastoral activities and the emerging agricultural societies also informs the myth of Cain and Abel.[118] The shepherd Abel made a more acceptable offering to the Lord than the farmer Cain, who consequently murdered his brother out of jealousy and thereby earned God's wrath – *your brother's blood cries out to Me from the ground! Therefore, you shall be more cursed than the ground, which opened its mouth to receive your brother's blood from your hand. If you till the soil, it shall no longer yield its strength to you.* Cain was then cast out although, just as in the case of Adam and Eve, God accompanied the banished. This indicates the inevitability of our acting against our own psychological interests and the universal opportunity for individual re-unification with God. There are clear congruencies here with Buddhist conceptions of the origin of agriculture as the source

---

[115] Miles (1997) Page 7
[116] Miles (1997) Page 34
[117] Genesis 3:14-19
[118] Genesis 4:1-16

of hoarding and of attachment to possessions. Agriculture is contrasted with an earlier golden age, as we will discuss in later chapters. Not unlike sustainable agriculture and Indian religions, the implication of Genesis that we are to be vegetarian in diet[119] (a point which seldom attracts the attention of modern Christian literalists) may similarly reflect the literate agricultural communities' dominance of the pastoralists at that time.

Rather than being indications of sustainable agriculture, these early references may reflect the importance of agriculture. For agricultural surpluses were essential to the continuing support of the priestly, learned and ruling classes, which defined ethical behaviour for the populace. Development of ethical rules does not seem to have been more or less effective in theistic communities than in the secular ethical approaches of polytheistic communities. However, Judaism's ascription of moral values to God placed them above the common human values of power, wealth, and pleasure, and in so doing provided a major developmental step in human self-understanding.[120] The detail of the 'laws' enumerated in the Pentateuch – the first five books of the Old Testament – is the fuel for the modern literalists' fires on which offenders are burned for 'crimes' as diverse as minority sexuality and unsustainable environmental actions, including mainstream agriculture. Such codes as The Book of the Covenant – which follows the Decalogue or the Ten Commandments – apply primarily to an agrarian community struggling with an emerging issue of property ownership.

To round off the two themes of this chapter, we may conclude that literal interpretations of scripture provide little of value unless their contextual intent is considered. In that

---

[119] Genesis 1: 29

[120] Miles (1997) Pages 110-111

case, many references that are relevant to sustainable agriculture and to most other details of ethics cannot conform to an unchangeable fundamental truth for each circumstance. We summarize these points as follows:

- Agriculture was the major occupation at the time the Christian scriptures were written, and hence references to it abound; a literal interpretation can seem relevant to agricultural sustainability in such forms as stewardship, land distribution and simplicity of lifestyles.
- When literal interpretations are tempered by the central consideration of the emergence of human self-consciousness, and the historical context of agriculture, then myths and detailed rules are understood to be more important to the development of social structure than to agriculture itself.
- Immutable scriptural references to sustainable agriculture separate from personal self-transformation are not readily found in the Christian scriptures.

We have considered the fundamental human fear of death as a source of the sustainability ethic, have searched for deep Christian insights that inform the intent of Christian teachings, and have considered of the errors of literality. All of this has focused at the individual level, the level on which essential religious teachings aim to effect personal transformation. To seek relationships between sustainable agriculture and Christianity necessarily involves the institution of the church and in particular its influential thinkers. This is the subject of the following chapter.

Chapter 5

Some Influencers of the Church: Prophets and Sustainable Agriculture

He who binds himself to a joy
doth a winged life destroy
he who kisses joy as it flies
lives in eternity's sunrise
Blake

The Christian church today has alienated its flock, and its multiple and sometimes conflicting global objectives easily offend the educated Westerner. Yet, it continues to attract and succour powerful, intellectual and spiritually insightful people. The influence of the latter type, the insightful persons, may seem minimal at any one time but over a longer period can be seen to resemble the prophets who spoke from a broad base of wisdom. While many may dismiss the church as a self-serving institution of little modern relevance, in its varied forms it continues to provide a foundation for social and spiritual vigour. It is also a barometer of Christian society's wellbeing. If the church is in a parlous state, so is our society. If the church condones flexible ethical values to suit current societal whims, it looses rather than gains influence.

The church formulates ethical statements and actions that affect everyday environmental interventions and agricultural sustainability. These formulations derive from diverse sources. Sometimes they project modern values back into scripture and other times they derive from ancient insights. This chapter considers some contemporary and historic trends in the church and finds remarkable consistency with the conclusions of other traditions as regards sustainable agriculture.

Christians often focus more on the New Testament than the Old, and often regard the latter as prophecy pointing to the former. Such an approach may be correct in certain allegorical and historical contexts of spiritual understanding, but must otherwise be relegated to the error of literality described in Chapter 4. The literal story of Jesus, for example, as expressed in the creeds is now an outdated myth. Yet, the use of myths is a powerful means of demonstrating complex matters and of unifying society. This is well known to great religious leaders and politicians alike. Understanding of the intent of 'the Jesus myth' today is benefiting from interaction with other global traditions and in so doing informing more of modern global needs. Examples abound in current USA literature about North American Indians, and may also be found in statements of modern Christians. For example, *there is only the flux of life, the passing show of existence ... just delight in life, delight in experience, delight in the way the world continually pours out and passes away. I want to shift away from the notion that only the eternal, only the unchanging is religiously valuable, towards delight in experience.*[121] So if variations in natural phenomena are inevitable, we should accept change rather than fret over secular definitions of sustainable agriculture.

Such a view is clearly in conflict with the modern capitalist ethic that is now almost inseparable from popular Christianity. The modern ethic takes a long-term perspective in saving for future rewards at the expense of *joy in life here and now*. Sustainability forms part of the capitalist-Christian ethic of planning for the future, and it is difficult to fault in the type of rational reasoning condoned by that society. Coupled with Western intellectuals' flirtation with an atheistic belief system – that is another belief system just as religion is a belief system

---

[121] Geering (2001)

– God becomes a problem for those who eschew the church yet avow Western capitalism. So the church continues to redefine itself without separating from its Western societal origins, and in so doing further obscures the central message of the fulfilled life in the Jesus story. Perhaps this confused situation explains why Eastern religions are increasingly invoked as a means of non-theistically portraying the fulfilled life. Thus is fulfilled Whitehead's prophecy: *The Buddha gave his doctrine to enlighten the world: Christ gave his life. It is for Christians to discern the doctrine. Perhaps in the end the most valuable part of the doctrine of the Buddha is its interpretation of his [Jesus] life.*[122]

Such comparisons do not compromise essential Christian values, although they do challenge church doctrine and literal belief. We may understand this simply by recognizing that the belief system of the church has varied across eras. The beliefs of Jeremiah, Augustine and Luther varied one from the other, but they were *all people of faith, great faith, and they belong to a tradition of faith.* And the faith referred to was a confidence of the personal development and fulfilment available through specific spiritual exercises, not through blind acceptance of a literal interpretation of an allegory. Evolving from the central beliefs of creeds through interaction with new global challenges may be seen as, in the spirit as Miles' analysis of the Old Testament, a continual refinement of the personality of God as a group of people became more aware of their own consciousness.[123]

God-consciousness thus becomes the experience of human consciousness. Within such a global historical perspective, there is no *transcendent authority apart from the universe itself, of which we are a part. Some scientists have said the universe is becoming conscious in us* – a thought remarkably similar to the evolution of consciousness described by

---

[122] Whitehead (1996)
[123] Miles (1997) Page 56

Griffiths[124] and discussed in a later chapter. In such a vision, we see Jesus' homely teachings of the kingdom of God as portraying ultimate harmony and universal peacefulness in everyday human settings. Apart from the distractions of treating references to heaven and so forth literally, they are also easily misunderstood in humanistic or pantheistic terms. But their essence is in fact congruent with that of great mystics from diverse traditions. This must cause us to consider the implications of the mystics in everyday spirituality as it relates to change - and change is the imagined enemy of sustainability.

If natural fluxes are accepted, then it would seem that Jesus and essential Christianity taught of the joy that may be experienced in change. This is the intent of Blake's citation with which this chapter opened. Without a huge leap in logic, we might thus say that agricultural sustainability can be a manifestation of that joy when change is part of our experience.

Is this acceptance of change being discussed in today's church? It seems it is, and to effect. To take some examples of reforming influences on the modern church, we may consider such figures as Spong, Cupitt, Tillich and Teilhard de Chardin. Their interpretations are both learned and socially aware, and of apparent appeal to the searching remnant of the modern church. Spong considers the essence of Christian faith as teaching us to equip ourselves to accept the insecurity of the world. And he calls the church's alternative approach of striving to create security 'modern Christian idolatry'. That idol worship is preferred by the majority who, unable to live with the insecurity of seeking God within themselves, seek refuge in literal definitions of God. The message is timely and derives from the writings of such diverse Christians as Aquinas and Tillich. In each case, God is increasingly

---

[124] Griffiths (1992)

understood more as an experience than a defined being[125] – an approach similar to the Buddhist conceptions of consciousness that we discuss in a later chapter.

The second modern prophet is Cupitt who seeks an essential Christianity by tracing the roots of religion to early agricultural civilizations with their cycles of feasts. For him, the separation of Western society from agricultural cycles has undermined the church. It is no longer clear to most Christians that the Easter story blends with the seasonal rebirth of spring as another metaphor for the spiritual rebirth of experiencing 'heaven on earth'. Nor is it clear to the majority that the Christmas story coincides with the ancient winter solstice rituals that acknowledge the symbolic banishment of the darkness by the birth of an ever growing light as a representation of the same personal enlightenment. When the stories are taken literally, they can easily loose their essential meaning in non-agricultural communities. Yet, as the Quakers show, it is possible to retain the essential Christian message even in our post-agricultural age. Among the Christian groups, the Quakers may be the closest to the early followers of Jesus, for they have retained an essence of individual relationship to God in a manner that again appears akin to Buddhism. This point has not escaped Cupitt who described Buddhism as *the strongest religion intellectually.*[126] A theology of the individual that looks beyond attempts to create a stable world ruled by law[127] represent the essential Christian stance that informs agricultural sustainability.

To be at home with change as part of the Christian message allows us to be *content to be part of the flux from which we came and into which we will return, which has produced us and which we also produce.*[128] Heaven is thus readily seen as a state

[125] Spong (1998)
[126] Cupitt (2000)
[127] Cupitt (1995) Page 149
[128] Cupitt (1995) Page 8

to which we may aspire here and now. When freed from the constraints of literal religion we effectively become one with all things, conscious or co-knowing (*con-* 'together' + *sci-* 'knowing'), part of nature. And when we see that everything is material, we know that nothing can be alien to us - we are always at home.

Being at home is therapeutic. In Greek, *therapeia* means 'a remedy' or 'a household servant', and the form *therapein* means 'household gods or idols'. These derivations proved the beautiful realization that a therapist is one who makes us feel at home by bringing the gods to attend where we abide. Such wholeness and wellbeing gives life to teachings of the immanence of the spirit. Yet this understanding is far from that of most of the church, an occurrence explained by a somewhat Freudian analogy of a son's relationship with his mother.

Cupitt uses this analogy to explain some unpopular aspects of traditional Christianity. He reasons that Christianity is like a man who falls in love. The woman whom he loves reminds him of his mother because she evokes the *kisses and cuddles of childhood*. He develops the possessive feeling that encourages the expression of demanding male tendencies. In the same manner, the patriarchal God is initially a capricious jealous God. His overwhelming ego demands attendance from an errant wife or child, cast as Israel. The ideal sinless woman who is the obedient and adoring mother desired by every man became God incarnate as Mary. But religious images of Mary never express physical contact or affection with Joseph. She is wholly a mother and never a wife. The baby Jesus held in her arms reflects the comfort sought by males. Even at death the crucified Jesus is repeatedly portrayed in Mary's arms.

From this perspective we see that the patriarchal God of the Old Testament, unable to elicit pure devotion from Israel, wins humans' love through showing his inner weakness. If we see that the whole Judeo-Christian tradition has centred on the male ego with the *fragile Omnipotent, the bossy Babe and the vulnerable Male* being the basis of Western civilization, then it is little wonder that it wages war, blames, rationalizes, and exploits so well.[129] Let us return to our wider theme for a moment for it is the politically-incorrectness of the aggressive patriarchal society that motivates some would-be sustainable-agriculturists. This is why agriculture is cast as aggressive and uncaring, thereby indicating an additional link to feminist more than religious thought, both of which can sometimes overlook natural law.

An inherent tension exists between humans and natural law, which is the source of our perennial actions of seeking and fleeing from God.[130] The paradox of fleeing from that universal, benevolent and benign God who confers the happiness and immortality that we desire eloquently shows that we create most of our own problems. We know this at the most fundamental level because we actually enjoy our sins. We will see later how this is the same message as that of Buddhist teachings that we can only be disappointed when we crave for something, even seeming virtues such as sustainability and gender equity. To understand this requires in the Christian terminology, being 'reborn in Christ', 'dwelling in the kingdom of God' and so on. It means abandoning false self-knowledge and being *willing to look into a deeper level of our being*, for as Tillich says, *in the depth is truth, and in the depth is hope; and in the depth is joy.*[131]

That joy exists everywhere, even among the clanging cymbals of the secular world. Yet we are deafened to the natural music of the universe and so fail to appreciate our unfulfilled need to reconcile with nature. This is what we 'fell' from in the story of Eden, and it is just one more way of expressing the reasons for suffering. Within these insights, the

---

[129] Cupitt (1995) Pages 79-80
[130] Tillich (1949) Pages 46-58
[131] Tillich (1949) Pages 59-70

sting of death[132] may be interpreted as the loss of eternity. And this is our loss of connection to nature. Redemption is salvation from the 'sin' of separation from nature, from God.[133]

So we may conclude that our attempts at immortality or selfish sustainability must fail, for our own purposes are not nature's. Our true salvation is nature's salvation, as portrayed in such images as Isaiah's leopards and kids lying down together[134], and as incorporated as the essence of oneness in the Christian sacrament. *Therefore, commune with nature! Become reconciled with nature after your estrangement from it. Listen to nature in quietness, and you will find its heart. It will sound forth in glory of its divine ground. It will sigh with us in the bondage of tragedy. It will speak of the indestructible hope of salvation.*[135] When we do not listen in quietness, we risk acting against our own salvation and harming other components of nature, which is the reason for the apocalyptic images of religion.

The apocalyptic messages of the modern environmentalist often echo those of a Hebrew prophet who is observed to *even in the greatest ecstasy, not forget the social group to which he belongs, and its unclear character which he cannot lose.*[136] Anathema to the establishment and the populace, such true prophets have readily been upstaged by false prophets. Are we academically titled experts who proffer human-managed agricultural sustainability only popular because our message is false? When we read that prophets *came to fulfil the law*[137] according to their differing gifts yet moved *by the same spirit,*[138] we may say, without hyperbole, that they saw a wider vision of reality than their peers and that the reality they saw is the same as that seen by seers of all times and in all places.

---

[132] I Corinthians 15:56
[133] Tillich (1949) Pages 170-173
[134] Isaiah 11:6
[135] Tillich (1949) Pages 82-91
[136] Tillich (1949) Page 96
[137] Matthew 5:17
[138] I Corinthians 12:1-11

Another modern prophet, Teilhard de Chardin interpreted this essential Christian message as a progressive growth to a higher form of consciousness, one that extends from the individual to embrace all of us. Similar to the visions of ancient mystics and even Jung's collective consciousness, this perspective links scientific knowledge to a divine 'milieu' imbued by an omnipresent God who manages evolution according to his divine nature, and ultimately reunites all within one perfect whole.

Living in that one perfect whole, we work in accordance with God's natural laws in contrast to our normal hermitic separation from them.[139] *In the divine milieu all elements of the universe touch each other by that which is most inward and ultimate in them.* They all share *the same reality … in their innermost being.*[140] *We cannot be fundamentally happy but in a personal unification with something Personal (with the Personality of the Whole) in the Whole. This is the ultimate call of what is termed love.*[141] This love contrasts with the conflicting urges of *selfhood, individualisation, and separation* on the one hand, and *escape from the loneliness of self into something bigger than self* on the other. Reluctant to give up selfhood, yet capable of participating in something much greater, it is usual for us to suppress our inner longings for reunion with that unspecified divine element that feels hauntingly familiar.[142] Here is the existential issue of sustainability - we say we want it but we are not willing to act on what we feel will produce it and will be universal benefit.

Our relationship with nature is also confirmed by the experiences of other mystics, who too transcended the linking of mind and heart.[143] While their messages often belie

---

[139] Teilhard de Chardin (1972).
[140] Teilhard de Chardin (1972) Page 114
[141] Teilhard de Chardin quoted in Happold (1991) Page 398
[142] Happold (1991) Page 40
[143] Berger (1914) Page 75

classification, analysts have nevertheless placed these into three categories; Pan-en-henic - all in One and that One in all, Pan-en-theistic - all in God and God in all, and Pan-theistic - a Deity or the divine contained in creation.[144] These mystical views provide the knowledge *that the cosmos, which to the self-conscious mind seems made up of dead matter, is in fact far otherwise – is in truth a living presence. [The mystic] sees that the life which is in man is eternal.*[145] God and eternity exist here and now.[146] Thus the Sermon on the Mount, for example, is interpreted in such words as: *If you are to live in [the world] successfully, you must conform with [its nature and the moral laws which govern it]. If you frame your actions in accordance with them you will be like a man who builds on a firm foundation; for you will be living in accordance with the real pattern of the universe ... you cannot fail.*[147]

We might say that sustainability seeks to reach that same place - of not failing. Success is obviously tied to our conforming to some natural order. Are such mystical communications relevant to sustainability? I think so, for the mystic crosses cultural mores to such an extent that Christian mystics often sound the same as those of the spiritual revolution around the middle of the last millennia BCE in India. And it is from that era that we learn of our place as part of nature. At that time, the early Indians' attempts to reconcile spirit and matter provided a foundation for an understanding that long pre-dated the Christian insights that we have discussed above. We examine such Indian insights in later chapters while in this chapter we allow ourselves the luxury of isolation from distant cultures, and return to the West's Hebraic and Greek foundations.

---

[144] Happold (1991) Page 43
[145] Bucke (1905) quoted in Happold (1991) Page 55
[146] Happold (1991) Page 87
[147] Happold (1991) Page 102

The historical orientation of the Hebrew scriptures as a revelation of spiritual realization contrasts with the Indian focus on the inner, non-phenomenal world. The Hebrew God directed history and required righteousness and justice of his people. Greek rationalism subsequently expanded philosophical concepts of the phenomenal world and blended them with the metaphysical components that we know as Platonic and Neoplatonic thought. From these elements emerged not only Christianity but Christian mysticism.[148]

The mystical existence of God within us, rather than a separate superego was *unto the Jews a stumbling block, and unto the Greeks foolishness.*[149] This means that it presented a theological barrier to a Hebrew monotheism that saw God as separate from us. And to the logical Greek mind, the immanence of a perfect God within an imperfect creation seemed irrational.[150] It was thus a paradox, a mystery - sometimes construed as all matter being an incarnation of the omnipresent spirit that we do not readily perceive.[151] Be that as it may, the simpler explanation of our use of the idea of God as a means of discussing our consciousness of ourselves may ultimately serve a wider purpose, and this is a message remarkably similar to other spiritual traditions.

Realizing these fundamental similarities is probably the greatest gift of the current global age. When we look beyond the self-serving emphases of traditions and their institutions, we can but wonder at the amazing congruity of the outcomes of unconnected persons applying consciousness to understand consciousness. A few of countless similarities include the Mahayana Buddhist ideal of the Bodhisattva, the mysticism of the Bhagavad-Gita, and the loving intermediary of Jesus

---

[148] Happold (1991) Pages 104-110
[149] 1 Corinthians 1:23
[150] Happold (1991) Page 110
[151] Happold (1991) Page 120

between suffering humans and God.[152] All relate to the ideal of reconciliation or unity. The Christian concept of cosmic unity is inferred in such verses as: *all things are of God, who hath reconciled us to Himself by Jesus Christ, and hath given to us the ministry of reconciliation; to wit, that God was in Christ, reconciling the world into himself;*[153] and *in Him should all fullness dwell ... to reconcile all things unto Himself; by Him I say, whether they be things in earth, or things in heaven.*[154]

When Westerners read such verses, they do well to remind themselves that they are conditioned by their tradition, which derives from Greek philosophy. And that tradition is oriented by its mechanistic worldview. Yet the Greek masters were much more than what we call philosophers today. Plato, born around 437 BCE, outlined a spiritual philosophy that supported his better known political writings. *It was his intense sense of the world of spirit which impelled him to strive to create on earth the sort of state in which the life of the spirit would be possible.*[155] The 'soul' recalls a long-forgotten unity corrupted by association with, or denied opportunity by, adverse physical circumstances.[156] Socrates has Plato say that *beauty is first of all eternal; it neither comes into being nor passes away, neither waxes nor wanes,* in sentiments that are daily echoed in the Mahayanan Buddhist Heart Sutra. Such understanding inspires moral behaviour in accord with the soul's 'remembrances' of true reality.[157]

However, the West has received an impoverished tradition. In the transition of Christianity into a politico-religious institution, many metaphysical Greek insights were ignored. As a further example, Plotinus (205-270) followed Platonic philosophy and postulated a universal soul present in

---

[152] Happold (1991) Page 186
[153] II Corinthians 5:19
[154] Colossians 1:20
[155] Happold (1991) Page 175
[156] Jowett's translation quoted in Happold (1991) Page 181
[157] Hamilton's translation quoted by Happold (1991) Pages 183-184

all plants, animals and humans. The soul determines their form and actions. He conceived the universe as one life-form composed of these parts. A universal force united the whole and was felt by all life, animate and inanimate, material and non-material.[158] Whether the West's reliance on rational Greek thought dates from the formative years of Christianity or the Renaissance, Western religion and science has lost this unified understanding. It no longer automatically feels what Plato called *remembrance of the soul in all.*

The Eastern Orthodox Church, on the other hand, has retained an openness to that life force, by practically accommodating mystical insight in its teachings. Developing separately from the Aristotelian thought that has shaped Catholic and hence Protestant theology, the Orthodox church has emphasized doctrines of unity (the Trinity) and grace, and fostered integration of dogma, theology and mysticism.[159] The Orthodox Church may well prove to be a bridge to other Oriental religious interpretations in the West's quest to interacting sustainably with other life.

In summary, we may glean the following main points:

- The insights of 'prophets' reveal a Christianity that seeks to enjoy the changeability of life, which is an expression of universal divine immanence, thereby suggesting that sustainable agriculture only exists within changeability.
- By seeing God as an experience rather than as a being, we become conscious of the immanent spirit and strive to act in keeping with natural law; on the other hand, emphasising rational or dogmatic approaches encourages us to act against such laws and defines sin – thus agriculture that contravenes nature is 'sin'.

---

[158] Happold (1991) Pages 201-210
[159] Happold (1991) Pages 218-227

- The traditional church has maintained a paternal orientation that emphasises control, which incorrectly defines human-controlled systems as sustainable.

We have considered Protestant and Catholic thinkers. Those who rebelled against the religious mainstream in the recent past also offer further explication of sustainability. This forms the subject of the next chapter.

Chapter 6

From Luther to Jung: Broadening the Insights

Love divine, all loves excelling
Joy of heaven, to earth come down
Fix in us Thy humble dwelling
All Thy faithful mercies crown
Wesley

An institutionalized and hierarchical church considers insights that conflict with its dogma to be irrelevant. If they gain any currency, they are heretical. Nevertheless, it is to such heretical and irrelevant insights that we often owe the ongoing revelation of Christianity. Their progressive unveilings may therefore offer us another perspective on sustainability. This chapter uses a further sampling of Christian insights in the same quest, ranging indiscriminately across the Protestant and Western Catholic traditions from Luther's time to Jung's.

Luther's insights were stimulated by the *Theologia Germanica*, a short manuscript perhaps written by a Teutonic Knight in the fourteenth century. Published by Luther in 1516, it became popular in the Reformation until the increasing rigidity of Protestantism eventually condemned it. The congruence of the Teuton's work with Eastern teachings is striking - consider the following: *sin is nought else but the creature turneth away from the unchangeable God and betaketh itself to the changeable; that is to say, that it turneth away from the Perfect, to 'that which is in part' and imperfect, and most often to itself.*[160] It also says that, *of all things that are, nothing is forbidden and nothing is contrary to God but one thing only: that is, self-will, or to will otherwise than as the Eternal will would have it. Remember this. For God saith to Adam, that is, to every man, 'Whatsoever thou art, or doest, or leavest undone, or whatever cometh to pass, is all lawful and not forbidden if it be not done from or according to thy will, but for the sake of and according to My will. But all that is done from thine own will is contrary to the Eternal will'.*[161] If this insight underpinned the Reformation, then the church of today has been able to suppress it in its common message. And as selfish actions are contrary to God's law, selfish attachment to sustaining something contravenes the essence of Christianity.

The sixteenth century mystic, St John of the Cross, spoke of a complete detachment and love that accompanied the experience of God in all things. God constitutes natural unity. *The soul is able to see, in that tranquil wisdom, how of all the creatures – not the higher creatures alone, but also the lower, according to that which each of them has received in itself from God – each one raises its voice in testimony to that which God is. She sees that each one after its manner exalts God, since it has God in itself according to its capacity; and thus all these voices make one voice of music, extolling the greatness of God and His marvellous knowledge and wisdom.*[162]

The same basic truth may be seen in the natural mysticism of the seventeenth century Englishman Traherne. Seeing human enjoyment as complete only in association with nature, he explains *the world is a mirror of infinite beauty, yet no man sees it. It is a Temple of Majesty, yet no man regards it. It is a region of Light and Peace, did not man disquiet it. It is the Paradise of God. It is more to man since he is fallen than it was before. It is the*

---

[160] Winkworth (2004) Chapter 2
[161] Winkworth (2004) Chapter 1
[162] St John of the Cross (1975) Stanzas XIV-XV

place of Angels and the Gate of Heaven.[163] While some may interpret this as simple pantheism, we may see it as an evocative elaboration of the Protestant mystic Boehme, who influenced Law. And it was Law's eighteenth century insights, shared as a Fellow of Emmanuel College Cambridge, that described our indwelling divine nature that constantly seeks to reunite us with its source, God. He uses, among other devices, the parable of the Prodigal Son which Christianity shares with Buddhism, to explain God as the mutual attraction of the divine in us and in all things.[164] We may also see it as the same as Tillich's later exposition on the theme of all nature groaning to reunite with God.[165]

This unity of all things (as described in Chapter 3) is significant for our consideration of sustainable agriculture. Practising agriculture within this wholeness - or expressed in another way, within natural processes and flows - defines the moral dimension of agriculture. But to understand the processes and flows of nature requires deep insight. Such revelations are the gift of but a small and usually uninfluential minority. Within this group, the practical approaches of post-Reformation nature mysticism may offer a useful framework.

One nature-mystic, the nineteenth century Jefferies described his experience of nature and soul without prior knowledge of other mystical writings. He advocated withdrawal from habitual worldly places and actions in order to keep the mind open to spiritual insight. Notwithstanding his isolation, his writings confirm eternity and immortality as universal aspects beyond all cycles and measures of time.[166] However, as he reminds us that our attempts to control anything cause us to forget that time is merely a convention, we are tempted to dismiss his insights as impractical.

---

[163] Traherne (1988)
[164] Law quoted in Happold (1991) Pages 376-383
[165] Romans 8:19
[166] Jefferies (1991) Chapters I-III

One practical application of these emerging insights came from the twentieth century prophet Emerson who linked meditation to everyday life through agricultural metaphor in such poetry as *the prayer of the farmer kneeling in his field to weed it.* He observed that while most of us seek health and wealth, a higher class lives for art, poetry, nature and science. That class is in turn surpassed by those who live for reality itself. The few who progress through the classes eventually *pitch [their] tent on this sacred volcanic isle of nature; [they do] not offer to build houses and barns thereon, reverencing the splendour of the God which [they see] bursting through each chink and cranny.*[167]

This transcendental vision conceived a unification of souls with nature, a *unity, that over-soul, within which every man's particular being is contained and made one with all other.* Emerson unwittingly commented on sustainability when he went on to say *the universe is fluid and volatile. Permanence is but a word of degrees.* His reaction against the mechanistic rationalist lifestyle that accompanied the Protestant expansion in the USA of his time caused him to observe that *logic is the procession of proportionate unfolding of the intuition; but its virtue is as a silent method; the moment it would appear as propositions, and have a separate value, it is worthless.*[168] Today, most of our sustainability research is based on propositions and separate values.

Society has travelled a different path from Emerson's and it strives to sustain its hard won and illusory victory over nature. We downgrade such emotions as being glad for the gladness of another and in fact act as if the *system is one of war, of an injurious superiority* - Emerson's words for market economics and competition. When he writes that *the cow is sacrificed to her bag, the ox to his sirloin ... stall feeding makes sperm-mills of the cattle, and converts the stable to a chemical*

---

[167] Emerson (1900) Page23, 54
[168] Emerson (1900) Pages 68, 75, 82

factory,[169] we may be sure that he would deem these modern infringements of animal welfare as unsustainable agriculture.

The rapidly developing industrial approach to agriculture in his day was an affront to the spiritually sensitized Emerson. He sought and saw a morality for humans in nature. He was able to ask *what is a farm, but a mute gospel?*[170] Yet he also noted (in words echoed by Tillich a century later) that unsustainable practices make us *as much strangers in nature, as we are aliens from God.* And such alienation from reality shapes all in our society, even modern education.

If *'the Laws, Divinity, Natural Science, Agriculture, Art, Trade, Letters – have their root in an invisible spiritual reality,* our educational systems and society at large have strayed from that reality. And it is here that we find that Emerson is read for his 'views, prose and poetry' rather than as a prophet. But like all prophets, Emerson was a product of his time – for if it was otherwise, we would have acted long ago. We note his anomalous support for the felling of previously impenetrable forests to create new agricultural land that could flourish through the application of science and so disprove Malthus' apocalyptic predictions. Rather than reconcile such a statement with his rejection of the reductionist science that fuels technology, which he describes as *hunting for life in graveyards,* we reduce him to one more opinionated writer. But as part of his society, he was advocating such actions within careful tending of new lands as a spiritual activity.[171]

The spiritual activity of agriculture was further noted in the commonality of agricultural metaphor across traditions. *In the Norse legend of our ancestors, Odin dwells in a fisher's hut, and patches a boat. In the Hindoo legends, Hari dwells a peasant among peasants. In the Greek legend, Apollo lodges with the shepherds of Admetus; and Jove liked to rusticate among the poor*

---

[169] Emerson (1900) Page 166, 254

[170] Emerson (1900) Page 320

[171] Emerson (1900) Pages 371, 442, 575

Ethiopians. So, in our history, Jesus is born in a barn and his twelve peers are fishermen. Tis the very principle of science that nature shows herself in the leasts; twas the maxim of Aristotle and Lucretius; and in modern times of Swedenborg and of Hahnemann.[172]

Emerson spoke on sustainability when he observed the common rejections of the transcendental message of Christianity, noting that *we delight in stability, and really are interested in nothing that ends.*[173] This defines the Western 'industry' of sustainability that at its worst disguises selfish ends by rhetoric of future generations, fellow creatures, and the intrinsic value of nature, and which is now embraced by the church. And such ethical issues sit uncomfortably with both the transcendental message of the church and its dogma. This discomfort provided a fertile ground for the West's understanding of itself through prophets steeped in the Christian tradition, such Jung.

Jung explained religion in terms of historical origins and the adjustments of culture. He saw that a spiritually disoriented Roman Empire had required the compensatory effect of a religion such as Christianity. Once elevated to the state religion, Christianity found itself in the position of needing on the one hand to rationalize its doctrines as a defence, and on the other in order to counter irrationality among its own adherents. This produced *that strange marriage of the original irrational Christian message with human reason, which is so characteristic of the Western mentality.*[174] Reason dominated and Christianity evolved as part of the modern technological age, but at a psychological cost.

Today primarily read for his psychological insights, Jung also interpreted our unconscious yearnings as reconciliation with God, other persons and nature. The

---

[172] Emerson (1900) Page 445

[173] Emerson (1900) Page 651

[174] Read (1970) Volume XI, Page 17

message is now familiar – religion *consists of the belief that there is an unseen order, and … our supreme good lies in harmoniously adjusting ourselves thereto.*[175] This is seen in the seamless association religion and ecology in India and Southeast Asia.[176] Today in the West this is often expressed in religious approaches to ecology that seek to reunite us with the divine.

Re-unification, the etymological origin of 'religion', may also be expressed in terms of healing the pervasive neuroses of society. In Christian terminology, we are re-unified when we accept ourselves and live according to our own 'inner calling', in the same manner that Jesus lived his life. This is the meaning of the *imitation of Christ.*[177] It differs from the usual rational interpretation of parables, stories and myths, which has sacrificed much of the non-rational reality that myths seek to convey. Jung considered the demise of myth a consequence of rationalism and a cause of modern psychological disorders. One of those disorders in turn is a constant sense of insecurity for which sustainability is grasped at as an answer.

To Jung, Christianity retains a salvific message that incidentally addresses the insecurity of change. He describes Christianity as *a story of how God created our world and found his creation to be good. In the course of time, and almost from the beginning, the world, in its freedom, became separated, alienated from its Creator. Yet God so loved the world that a particular time, now long ago, he sent his son, not to condemn, but to redeem the world through the power of love. In our state of separation (sin), we killed the very one he sent. God's love, however, was not thwarted. The story goes onto say that God demonstrated the power of love by using that death as a means of our salvation, raising the one he sent from the grave, empowering us with his own holy spirit, and offering*

---

[175] James (1994) Page 27
[176] Gosling (2001)
[177] Clift (1983) Page 72

us also a new life as participants in his new creation, the church, the body of Christ's resurrection.[178]

Thus the Christian story is a concept; the 'sending' of God's son and the 'exchange' of his life for human 'sin' portray the redemption of those persons who are reconciled to reality. The story is about reconciliation and salvation. Through discussion of myth, Jung revealed the role of symbolism in Christianity and so provided interpretations of ancient rituals as part of a universal psychological process. Such symbolism may be seen in the ritual of the Eucharist in terms of unification of substance, impermanence and inter-relationships.

So why do we miss the application of the essential teachings to sustainability? It is because dogma and creed have become so codified over the centuries that they have lost their original experiential element even though they have retained the symbols important to our unconscious. Deeply embedded in the Western collective unconscious, this symbolism is more powerful on a person from within that culture than on one who adopts a foreign culture.[179] In this manner, Western culture has formed Western minds such that the same symbols evoke consistent reactions. The same mechanism may be noted within all cultures, but here we are concerned primarily with Christianity.

In the Christian context, Jung's struggle to accommodate his experience of God described in his *Answer to Job* revealed a God that is both good and bad. The usual lop-sided God that is only good depreciated spiritual matters in parallel with appreciation of the physical world, which may have been what Nietzsche was getting at in his *Beyond Good and Evil*. Tentatively concluding that God *is only about three-quarters good*, Jung added the power of evil (symbolized in the

---

[178] Clift (1983) Pages 88-89
[179] Clift (1983) Pages 90-91

God-created serpent of Eden) to the Trinity. From his own and patients' psychological experiences, he noted the congruence of this Quaternity with such other insights as Taoism, the four sons of Horus in ancient Egypt, the four Evangelists, and even the design of rose windows.[180]

Myth also allows us to interpret the historical traditions of Judaism.[181] God in fact needs us. God is made in our image. God, after having created all things, was only able to find fulfilment through that one created form that has the freedom to reject his creative force of love.[182] God may be seen to draw us to him through his love expressed in creation - a love that exceeds all other loves, is unbounded and is an evocation to wholeness, as expressed in the lines heading this chapter, *Love Divine, All Loves Excelling*.[183]

The 'new creation' that is 'perfectly restored in thee' referred to in that hymn is the 'unified psyche' of Jung. It may be understood as a union of opposites that is a psychological healing - an unification with ourselves and the natural world. Thus, *taking up the cross of Christ*[184] becomes the first step of a reconciliation process towards maturity, characterized by internal misgivings. Reconciling opposites is the true symbolic intent of the Cross, it is just a matter of translating the Christian language. *For instance, instead of using the term God you say 'unconscious', instead of Christ 'self', instead of incarnation 'integration of the unconscious', instead of salvation or redemption 'individuation', instead of crucifixion or sacrifice on the cross 'realization of the four functions' or of 'wholeness'*.[185] A yearning for the calm of psychic unification fuels our fixation with sustainability.

---

[180] Clift (1983) Pages 131, 137
[181] Miles (1997)
[182] Tillich (1976) Page 422
[183] Charles Wesley in Methodist Hymn Book: Hymn 426
[184] Luke 9:23
[185] Read (1970) Page 153

Merton expressed our yearning for unity in the following way – *the world itself is no problem, but we're a problem to ourselves because we are alienated from ourselves, and this alienation is due precisely to an inveterate habit of division by which we break reality into pieces and then ponder why, after we have manipulated the pieces as they fall apart, we find ourselves out of touch with life, with reality, with the world and most of us with ourselves.*[186] This speaks not only of reductionism, but of our essential craving for stability – for sustainability, even though it is screened by the diversity, novelty and inventions of modern lifestyles.

Dulled by lifestyles of distraction, we are intellectually and spiritually asleep, unaware of the flash of *intuition by which multiplicity is suddenly comprehended as basically one – penetrated through and through by the logos, the divine fire.*[187] So we can see from this perspective that sustainability is based on a deluded conception of reality. The points of this discussion may be summarized in the following:

- Since the Reformation, insistence on rationality has diluted the Christian message and made many insights about our unity with an unseen order into potential heresies; yet this very fact can assist our understanding of sustainability as the ill-conceived self-will that has been defined in Christian insights as sin.
- Ex-church insights have commonly described a divine indwelling that permeates all nature and its flows, suggesting that agriculture may only be sustainable when practiced within those flows, which may be only understood by the wise.
- We have rejected the myths and symbols that inform us of change and our role and the healthy unification of our

[186] Cunningham (1992) Page 387
[187] Cunningham (1992) Page 287

> psyches lies in accepting the integrity and impermanence of all things, which is the first premise of sustainability.

If there are prophets of agricultural sustainability - as we might expect to find in a subject that often resorts to religious language and unsupported statements of belief - then we must count those mentioned in this and the previous chapter and their peers, for they have preached the integrity of all nature. At this stage in Western self-understanding, a self-preoccupation has been turned into a rational approach to understanding the psyche, and it is in this modern milieu that one should expect to regain respect for ancient insights. It is also a field that is finding affinity with Eastern traditions, and that in itself is causing us to consider Oriental insights more generously. The next chapter introduces some aspects of congruence between Western and Eastern traditions.

Chapter 7

West Meets East: The Salvation of Agriculture

The reality is that all religious truths come from an original experience, that of the seer, the prophet, the saint. The experience always has to be interpreted in the light of rational, conceptual thought.

Griffiths

Modern communication offers us the gift of understanding our global similarities and the artificiality of separating one religion from another. In pre-agricultural times until the evolution of complex language our ancestors may well have lived without sophisticated awareness of past or future. With the abstraction of language, they learned to communicate without pointing and touching, and were eventually able to develop such concepts as remembrance of the past and anticipation of the future. This probably occurred before the time that human economies shifted from hunting to agriculture. Yet in the knowledge handed down to us, history, the afterlife, and agriculture were triplets born.

Hunters are romantically said to have lived day to day, eating as they hunted food. They lived in the present. Agriculture, on the other hand, requires more planning for the sowing of crops, for harvest, for comparing harvests across years, and for learning the complexities of interfering with natural systems. Of course, both hunting and agriculture require planning. The point is that humans gradually ceased to be moulded only by the environment and began to mould themselves and their environment. This evolutionary conception of agriculture varies from the Christian misconception that agriculture was a 'religious injunction' against the unruly and unsophisticated herders and their nomadic lifestyles.[188] In both agriculture and herding communities, a religious dimension was integral with everyday activities.

The psychological and religious significance of the development of language is easily underestimated. Our forebears probably did not consider trees as either objects for investigation or for unconsidered use, but rather more like 'nymphs' within an overall conception of a living cosmos.[189] With agriculture, storage of food for the future became common and led to exchange and eventually to the introduction of currencies - precursors of today's spiritless 'commodities'. Surpluses of food also created the civilizations of great river valleys of the Euphrates, Nile, Indus and Yellow rivers, and freed a portion of the populace to engage in other pursuits - including writing. Good harvests were critical and propitiation of gods to favour good harvests was a primary focus of each developing civilization. Sacrifices became more demanding over time, eventually including human sacrifice. This contrasts with the pre-language social arrangements where killing and warfare are romantically, and unrealistically, held to have been unknown.

Logical thinking created a conceptual revolution that led to wider communication and deeper investigation. It also introduced the power to control ourselves and the external world. It became logical to substitute sacrifices of one's tribal fellows with captured troops, for example. The next major development was awareness of consciousness.

Awareness of consciousness is clear in the great period of Indian spiritual development, which began in the second

[188] Emerson (1900) Page 10
[189] Griffiths (1992) Page 38

millennium BCE. The pre-Vedic period may have been a matriarchal society that worshiped a mother goddess in a manner similar to Egypt and Mesopotamia - the origins of the god Shiva lie in this early period. The Aryan influence from southern Russia impacted on Greek, Latin, Celtic, Persian as well as Indian thought and produced the major development - the *Vedas*. The *Vedas* are among the most ancient poetry in the world and reflect an extant desire to unify all gods. In the Vedic conception, three components of the world - the physical, psychological and spiritual - comprise a unified and balanced whole. The psychological or spiritual sickness of modern society introduced in the previous chapter can be interpreted as resulting from its overemphasis on the physical or material more than the spiritual dimension. By contrast, the integrated vision of early communities seems to have produced a sort of sustainable agriculture, one that lasted for a millennium. It is this line of thought that produces today's romantized view of third world farmers, an aspect we will investigate in a later chapter.

According to the Vedas, a fulfilling life was one lived in harmony with the cosmos and its rhythms. Selfishly taking more than necessary is contrary to the natural order. This is clear in the later *Bhagavad-Gita,* which considers stealing to be the taking of food without acknowledging the natural order by making an appropriate sacrifice or thanks. By 800 BCE the *Aranyakas,* or forest books, mark a further development with the external rituals and myths evolving into a search for the inner self. The sacrificial fire became an internal fire to which fears, sins, thoughts and desires might be offered. This shift in consciousness is evident in the *Upanishads,* written around the time of the Buddha. Greek philosophers of the time such as Heraclitus were similarly developing the consciousness of a transcendent god. This period produced the major religious systems of the world - each as an evolution of self-understanding, including the Hebrew branch onto which Christianity is grafted.

The Hebrew God was separate from the world. The universe was an output of God's work rather than, as in India, being the mind itself, albeit often a god's mind. Understanding the immanent presence described as God in Judaism developed slowly. Nevertheless, the unity of creation is indicated where the creation story describes every tree as *pleasant to the sight and good to eat*[190] and in God walking *in the garden in the cool of the day*.[191] The arrival of the serpent that disrupted this harmony symbolized our falling away from that natural unity. We were described as being seduced by our lower nature, which became known as 'sin'. The historical conflict between agriculturists and pastoralists told in the story of Cain and Abel[192] links sin to the expansion of agriculture. The agriculturists were the Babylonians and Egyptians that pestered the Israelite pastoralists.

The danger of selfishness accompanies agriculture, and remains embedded in our anguish over agricultural sustainability. The story of the Tower of Babel, built *up to heaven*,[193] depicts in architectural metaphor the same dominance of the physical world over the spiritual. But it was not sustainable agriculture that produced the evils of Cain or the arrogance of Babylon, but an attitude that fractured understanding the integrity of all things. Re-uniting is the salvation theme of the Judeo-Christian and other traditions.

The repetitious Old Testament theme of reconciliation with God corresponds to our fundamental desire for re-unification with nature. We 'fell' by turning away from the spirit and focusing on ourselves, yet the spirit moves to reconcile us with God to restore us to the paradisiacal state

---

190 Genesis 2:9
191 Genesis 3:8
192 Genesis 4:2
193 Genesis 11:4

referred to in Isaiah. This is interpreted by Christians as a prediction of the arrival of Christ,[194] a product of metaphorical genius that is so often misunderstood by literal interpretations. The same error of literal interpretation of Judeo-Christian myths about agriculture overlooks the association of selfish flouting of natural flows with unsustainable outcomes - the wide assumption of a duty to use the earth for our own ends derives from such literal understanding of the creation myths.

The Genesis instruction to *fill the earth and subdue it*[195] has licensed widespread environmental violence. It forms part of a replacement of the interdependent religions of agriculturists with a hierarchical religions of aggressive pastoralists. This licence is a repetitive theme in the Old Testament. God instructs Israel to utterly destroy and otherwise dominate other peoples, thus defining the common theme of violence of the Semitic religions of Judaism, Christianity and Islam.[196] Yet this mandate to subdue betrays a potential higher consciousness that is explained more clearly in later scriptures of more limited social impact. For example – *you shall no longer be forsaken nor shall your land be called Desolate; but you shall be called My Delight and your land Married, for the Lord delights in you and your land shall be married*[197] indicates reconciliation with nature.

The symbol of one's land is also powerfully used to similar effect in the New Testament. This is beyond misinterpretation in such versus as, *if they had been thinking of the country they had left, they would have had opportunity to return. Instead, they were longing for a better country – a heavenly one.*[198] Seeking a homeland is a metaphor for a concept elsewhere called 'the kingdom of God' and other similar references. The

---

194 Isaiah 51:3, 53:12
195 Genesis 1:28
196 Griffiths (1992) Page 87
197 Isaiah 62:4
198 Hebrews 11:14-16

same message of reunification is indicated from Romans - *all creation groaning in travail* for reunification[199] to Revelations - *I saw a new heaven and a new earth; for the first heaven and the first earth have passed away and the sea was no more.*[200] So, the Christian message of reconciliation with nature is not merely an ecological vision but a complete self-transformation, one that reforms our attitudes from partial and controlling to living in the unity of all nature.

We may see the perfect unity depicted in the Eden myth as an undifferentiated consciousness and the Fall as our failure to listen to our immanent spirit. Separation from the unified whole of creation produces the social divisions of individuality. Divisions between genders occur, then between tribes, then between individuals, and eventually between what we think is our self and our own consciousness. This is what Paul was expressing in his contrast between the *psuchikos* - the 'soul or ego man' who focuses on himself, and the *pneumatikos* - the 'spiritual man' who remains at one with all things.[201]

When we are not at one with the spirit of nature in our agriculture, *the earth will bring forth thorns and thistles.*[202] This integrated vision contrasts with the separation required to use language. Rational minds require dualism to function and our science, philosophy and theology are built on it. *Our normal way of perceiving the universe around us is in terms of an inner and outer world.*[203] A united spirit is the higher states of consciousness referred to in Buddhism, the Brahman-Atman referred to in the Upanishads, the Al Haqq referred to by Sufi mystics, and the kingdom of heaven in the Christian gospels. We will return to this theme in later chapters as it is critical to our discussion of sustainability, as well as to sound

[199] Romans 8:21-23
[200] Revelations 21:1
[201] I Corinthians 2:14-15
[202] Genesis 3:18
[203] Griffiths (1992) Page 100

understanding at all levels from the personal to international development assistance.

These are not new thoughts.[204] *There are indications that the Greeks were influenced by Indian thought. Pythagoras lived in the fifth or sixth century before Christ, around the time of the Upanishads and the Buddha. It is said that he had been to India and there is little doubt that his works show Indian influence. He believed in reincarnation for instance, when the doctrine was rare in Greece, and he practiced vegetarianism and taught the practice of silence in an organized community – all elements that seem to derive from Hindu and Buddhist traditions. Plato himself was profoundly influenced by Pythagoras and inherited this mystical tradition. But again Plato, in his written works, always used the language of 'rational analytical thought', though there is obviously behind it a mystical vision which comes out especially in some of his letters. There is certainly a tradition of some hidden mystical wisdom in Plato, but what stands out in his writings is the highly rational intellectual character.*[205]

Greek influence is evident in the Christian concept of salvation as an emerging consciousness that conforms to *Upanishad* and Buddhist teachings. It is also another indication of theological and philosophical interaction between the Mediterranean regions and India. Likewise, the inter-relationships of the Trinity can represent the inter-relationships in nature in a unity understood by some deep ecologists. The complete knowing of this unity is what spiritual writings call 'love' – caring for everything as if it is part of us.[206]

In such insights East and West have ever met. Universal truths have informed spiritual development, even if cultural differences fostered division. When we understand the West's religious heritage, we see that its emphasis on

---

[204] Falvey (2002)
[205] Griffiths (1992) Pages 230-231
[206] Griffiths (1992) Page 234-239, 254

rational thought has locked it into pre-conceptions of what sustainability must be. This orientation defines sustainability as meeting our desires before those of all others. And that definition conflicts with the sustainability that can be derived from the spiritual intent of the West's primary espoused religion.

There is much on which to reflect in the West's rediscovery of its Eastern heritage. It may be summarized the following:

- Congruence of spiritual insights across major religions, often subverted by religio-cultural differences, indicates our potential for higher forms of consciousness, allusions to which are misinterpreted in our usual rational discourse.
- As the ancients progressively described their own consciousness in scriptures, they realized the ideal of balance between physical, psychological and spiritual development - a balance that is biased by modern emphasis on the physical.
- Judeo-Christian and other myths that associate agriculture with evils have the intent of highlighting motivations as the issue, not agriculture; they suggest that we cannot expect to sustain agriculture or anything else if we serve impure ends.

Most interpretations of sustainable agriculture are either of two things: (1) a rational if not realistic desire to maintain things the way we crave as a form of agricultural salvation, or (2) a misinterpretation of the mythological and allegorical intentions that convey a spiritual message of our oneness with all things. One such misinterpretation, often aligned with opinions about ecological and agricultural sustainability, is pantheism. This tendency to project our spirituality into all things is treated in the following chapter.

Chapter 8

Pantheistic Agriculture: Investing the Gods in Agriculture

Whoso reveres me as abiding in all things,
adopting the belief in oneness,
though abiding in any possible condition,
that disciplined man abides in Me.
Bhagavad-Gita

Pantheism sees the divine in nature. This is quite different from the concept of Buddha-nature in all things as used as a didactic device in some branches of Buddhism, as discussed in later chapters. It also differs from the Gita's lines above, which exist within an early conception of interrelatedness and responsibility to act within it. Nevertheless, modern interpretations of pantheism evoke powerful images that can engender sensitivity to nature that can seem unworthy of criticism. Pantheism pervades Western romanticism of the peasants in poor countries and their various nature spirits. And because such traditional agricultural systems are often claimed to be sustainable, a pantheistic association with sustainability develops. The search for sustainable agriculture can indeed be informed by understanding low-input traditional agriculture and its human interactions with nature, though it does not offer the answer in itself as we consider later in this journey. In this chapter, we concentrate on the enduring appeal of pantheism itself, and consider its relevance to sustainable agriculture.

Our cultural assimilation of pantheism may be readily seen in a comparison of such symbols as World Mother,

Mother Earth, the world as a dance, play, theatre, or even as a mechanical work. The World Mother symbols offer the feeling of being comfortable and safe in a warm amniotic world that offers no fears and all sustenance. This is often interpreted to mean that spirit is immanent in everything. 'Mother Earth', on the other hand, places us *on the earth* rather than *in the earth* (mother), and thereby raises questions of hierarchy, which we answer by placing ourselves above nature. Dance, play and theatre are metaphors for interaction and reunification with the immanent spirit. However, the conception of the world as a machine has the opposite effect of suggesting a world devoid of spirit.

We can now view these symbols as a progression from maternal to paternal. This is reflected in Greek myths where the Mother cults of Delphi were replaced by that of Apollo, a shift strengthened by monotheism with its patriarchal hierarchy. Messianism is possibly a reaction to the repression of maternalism. As the central message of all scriptures is self-transformation to see natural unity, it is relayed through both matriarchal and patriarchal models, often through metaphors of siblinghood and friendship. However, there seems to be a difference; the patriarchal and mechanical culture of the West makes appreciation of an ecological model more foreign than it would be to a cooperative community imbued with matriarchal symbols.[207] Modern Western interest in pantheism is not necessarily a continuation of a past maternal model, but does represent a yearning for something more than the mechanistic model can deliver.

Modern Pantheism draws support from all religions and results in *'scientific' pantheism*. This teaches that *the universe and nature are divine*. It purports to align *religion and science, and concern for humans with concern for nature*. It also claims to explain *life after death, and [to provide] the most solid*

---

[207] Moltmann (1985)

basis for environmental ethics [yet] requires no faith other than common sense, no revelation other than open eyes and a mind open to evidence, no guru other than [one's] own self.[208] However, it is not as a formal belief that pantheism is most prevalent, but as a worldview of literate and secular persons who seek an acceptable cosmology separate from church dogma, and who find pantheism compatible with their reading of the classics.

It is claimed that most of the early philosophers considered some unspecified and unseen essence to imbue all life and all things, which was recycled by life and death.[209] According to Thales, the essential substance was water with all other matter composed of deities,[210] while according to Anaximenes it was air. To Heraclitus it was fire, and to Anaximander it was an ethereal *apeiron* that existed before the gods. Anaximander (611-546 BCE) considered *that the infinite is the divine, for it is immortal and indestructible,* and that the parts undergo change while the whole is changeless and has no beginning or end.[211]

Stoic belief that the universe is animate, rational and possessed of a soul similarly suits the doctrine of pantheism. Zeno of Cittium (300-260 BCE) placed man in the role of accepting destiny and living according to nature unquestioningly because all things are imbued with the will of the universe. God was conceived as the cosmos (a Stoic term) and as the culmination of all gods and identical with a sentient being which is animate and rational.[212] *Man himself has come to be in order to contemplate and imitate the world. But the world, since it embraces everything and there is nothing which is not included in it, is perfect from every point of view.*[213]

---

[208] Harrison (1999)
[209] Ross (1924)
[210] Lawson-Tancred (1986)
[211] Waterfield (1996)
[212] Petersson (1920)
[213] Cicero (1985)

Marcus Aurelius (reigned 161-180) accepted the Stoic belief of an intelligent universe with a soul, together with the fatalistic conclusions that the philosophy implies. In his *Meditations* he advises to *constantly regard the universe as one living being, having one substance and one soul; and observe how all things have reference to one perception, the perception of this one living being; and how all things act with one movement; and how all things are the cooperating causes of all things which exist; observe too the continuous spinning of the thread and the contexture of the web … all things are implicated with one another, and the bond is holy; and there is hardly anything unconnected with any other things. For things have been co-ordinated, and they combine to make up the same universe. For there is one universe made up of all things, and one god who pervades all things, and one substance, and one law, and one reason.*[214]

Thus the universe changes constantly to create new things, and our striving against change is futile. This realization caused Marcus Aurelius to advocate the laying aside of carelessness, unreasonable action, hypocrisy, self-love, and discontent with one's lot, in a manner suggestive of the Buddhist causative arguments. Thus *if you work at that which is before you, following right reason seriously, vigorously, calmly without allowing anything else to distract you, but keeping your divine part pure, as if you might be bound to give it back immediately; if you hold to this, expecting nothing, fearing nothing, but satisfied with your present activity according to nature [you will be happy]. And there is no man who is able to prevent this. … Every part of me will be reduced by change into some part of the universe, and that again will change into another part of the universe. and so on for ever. And by consequence of such a change I too exist, and those who begot me, and so on forever in the other direction.*[215]

So already at this time it was known that sustainability cannot involve *striving against nature* or acting with an impure

[214] Hadas (1960) Sections 4.40, 7.9
[215] Hadas (1960) Sections 3.12, 5.13

attitude. By the time of Plotinus (died 270), an impersonal, infinite, eternal, unified, and omnipresent god had been conceived[216] who had no thought, knowledge or movement, which was like a limited form of the Tao. Describing out-of-body experiences in which he saw the *higher soul* extending almost into plants, Plotinus assisted our understanding of the interdependent relationships of the seer with the seen[217] - although his visions tempt some to literal pantheistic interpretations stripped of their more likely intent of transcendence.

To briefly consider an Eastern tradition that is also manipulated to serve modern pantheism, we can look to the Tao in its lines:

Being one with Nature, he is in accord with the Tao.
Being in accord with the Tao, he is everlasting.[218]

and

The universe and I exist together and all things and I are one

and

He who regards all things as one is a companion of Nature.[219]

But just as the Tao was interpreted politically by Confucians as preaching inactivity and social irresponsibility, so the pragmatist dominates the spiritual in modern society. Perhaps this explains the modern subversive appeal of pantheism over traditional Christianity. Of course, in disparaging the Tao as inactive and irresponsible, the Confucians were judging the non-rational in rational terms, an omission later perpetrated by modern pantheists.

Within Western culture, it is easy to see the pantheistic appeal of such words:

Whither shall I go from thy spirit?
Or whither shall I fall from thy presence?

---

[216] Gerson (1996)
[217] Gregory (1991)
[218] Fung-Yu Lan (1952)
[219] Legge (1962)

If I ascend to heaven, thou art there!
If I make my bed in Sheol, thou art there!
If I take the wings of the morning and dwell in the uttermost parts of the sea,
even there thy hand shall lead me,
and thy right hand shall hold me.'[220]

But this is a poetic expression of rising consciousness of personal spirituality and is, in any case, far from the final word of the Judeo-Christian tradition that continued to evolve for centuries after such Psalms were written.

Christianity understands the mystical union that is all things as transformational in the same way as other great religions, such as in Islam. The Koran records *wheresoever you turn, there is the face of God.*[221] The New Testament epistles record with the same intent that *He is before all things and in him all things hold together*[222] and *Know ye not that ye are the temple of God, and that the spirit of God dwelleth within you? If any man defile the temple of God, him shall God destroy; for the temple of God is holy, which temple ye are.*[223] Also in the Gospel of Thomas, Jesus says *cleave the wood, and I am there; lift the stone, and you will find me there.*[224] But no such references are limited to pantheistic interpretations. Neither do they speak directly about sustainability, no matter how much we might like them to do so.

Pantheism as a modern religion appropriates whomever it desires as its prophets. One can claim pantheism in the insights of Hildegard of Bingen (born 1098) who understood the Holy Spirit to be present in all things,[225] of Aquinas (born 1224-5) who reasoned that God must be present in all things yet also separate, of Eckhart (born 1260) who saw

---

[220] Psalm 139:7-10
[221] Koran II.115
[222] Colossians 1:17
[223] 1 Corinthians 3:16-17
[224] Gospel of Thomas 77b
[225] Flanagan (1989)

man's unity with God, of Spinoza (born 1632) who concluded that God exists beyond the sensed world,[226] and of Rousseau (born 1712) who revelled in nature and disdained the church. Coleridge, Wordsworth, Keats, and Shelley speak of a universal spirit, as does Blake in his oft quoted lines:[227]

To see a World in a grain of sand
And Heaven in a wild flower,
Hold Infinity in the palm of your hand,
And Eternity in an hour.

Pantheism is, in the main, a beneficial platform for environmental awareness. It provides a useful contrast with the environmentally destructive image of Christianity. Pantheism views nature to be sacred and thus worthy of preservation in terms of its beauty, diversity, and health, and encourages us to wider contact with nature. It has been represented both as a plausible religion in its own right and a belief system that can be practised in parallel with other religions. Pantheism also differs from the Gaia hypothesis,[228] which is a useful conception of the interconnectedness of all things without a belief structure - although it too seems to be being taken literally as a belief system by some. But while pantheism highlights our relationship with our surroundings, it falls short of our more critical relationship with our own consciousness. It can even be a stumbling-block when the examples it choses for its verification are misinterpretations of metaphors and allegories for the fruits of a higher consciousness.

Regardless of its limitations in religious terms, pantheism remains a useful device to consider the divine in all things. In agriculture it encourages care for land, soil, rocks, insects, plants, animals, and ourselves - and surely this can

---

[226] Elwes (1989)
[227] Blake (1996)
[228] Lovelock (2001)

only be good. But in missing the central message of the sages and scriptures, modern pantheism limits us to the small and uncertain gains of rational processes uninformed by significant spiritual development. It assists sustainable agriculture in the form of making the best out of a bad job.

We might summarize this as follows:

- In considering the divine in all nature, pantheism encourages a maternal and interrelated conception of life that fosters positive emotions toward the environment.
- In the story of human realization of consciousness, rational interpretations coupled with domineering paternal religious symbols have produced a reactionary seeking of maternal metaphors, which can support pantheism as an end in itself.
- Even though the classics and scriptures confirm the message of transcendence more than pantheistic beliefs, the respectful practice of agriculture as if all parts of it are imbued with one spirit is more supportive of sustainability than is a mechanistic attitude.

So while pantheism might be a step toward sustainable agriculture, it favours literalistic interpretations of scriptures that have other intentions. Its popularity is enhanced by our general reliance on the same rationality that has caused much of what would have otherwise been explained through religion to become the province of other philosophical fields. Such philosophical fields as ethics and the evolution of rights might thus be seen as part of the West's modern religion. This is introduced in the following chapter.

Chapter 9

Agricultural Philosophy and Rights: From Natural Rights to Rights for Nature

The fairest thing we can experience is the mysterious. It is the fundamental emotion which stands at the cradle of true art and true science. He who knows it not and can no longer wonder, no longer feel amazement, is as good as dead, a snuffed-out candle. It was the experience of mystery – even if mixed with fear – that engendered religion. A knowledge of the existence of something we cannot penetrate, of the manifestations of the profoundest reason and the most radiant beauty, which are only accessible to our reason in their most elementary forms – it is this knowledge and this emotion that constitute the truly religious attitude; in this sense, and in this alone, I am a religious man.

Einstein[229]

In this expression of religion, Einstein provides the unifying context of wonder to science, art, philosophy and religion. Without mentioning it, he also stimulates us to recall a past when the arts included the natural sciences, and when philosophy was integral with religion. It recalls the spirit of the Greek philosopher who lived his philosophy in a stratified society. In contrast, we can see modern philosophy as a means for consolidating disparate views to support revolutionary social change, such as the rejection of slavery or gender exploitation.

Notwithstanding Einstein's broad definition of being religious, it is easy to emphasize the rational over the

---

[229] Einstein (1996)

experiential in Western societies. In addition, Western ethics are today influenced by social permission. While some may see a decline in the influence of Christianity, the society is simply a wider context for decision-making that itself is a product of historical changes in ethical values. Religiosity is far wider than institutionally sanctioned actions, and can be appreciated through those social historians styled as philosophers. From these persons, we might expect to find a 'religious' understanding of sustainability. And these modern philosophers are building on the 'major spiritual tradition of the West', seen by some as Neoplatonism,[230] especially in the last four centuries.

Since the Reformation, secular expressions of spiritual values in the arts, science, and philosophy have served us well – with science exploring the external world while literature, music and the fine arts explored the inner world. Psychology integrated the two worlds to help understanding of personal consciousness. So it is from rational interpretations of the collective insights of these pursuits that we arrive at the modern shorthand of many ethical questions – rights. That individual humans have rights is almost axiomatic to Westerners; now the concept is being extended to animals or even nature in general.

At this point it is worthwhile reminding ourselves that 'rights' in the legalistic Western sense do not necessarily have any tangible basis. The concept of rights is a tool in the service of a political goal of equality, which itself is an idyllic conception. And just as tools can be misused or ascribed a mystical value, so has been the concept of rights. I am not simply appealing for rights to be re-linked with responsibilities – though that is a good first step. I am trying to maintain an open perspective on our limited appreciation of reality. That open perspective allows us to see, for example,

[230] Sangharakshita (1994) Page 17

that the concept of rights even when attached to responsibilities, is but a partial response to deep insights about nature, many of which are clearer in the non-theistic Eastern tradition of Buddhism.

Without pre-empting a later discussion, we may usefully note that the central Buddhist understanding of co-dependence relates treatment of all other things, living or not, to each other. It thus raises our perspective to a level of universal inter-relationships. Buddhist understanding can go a step further and reveal Western attachments to 'rights' as a source of angst when they are seen to be an individual's property. Asserting 'my right' to free choice/equal treatment/free speech/health care/safe streets and so on is accepted in many Western contexts as appropriate social behaviour and obviously produces stress. Rights can be divisive and can be more a source of conflict than a permanent solution in many cases. Notwithstanding this insight, the pervasiveness of the concept of rights makes the concept another tool in our examination of sustainability. So, we return to consider the rights of nature.

The history of the rights of nature that culminates in modern environmental ethics has been comprehensively traced through changing attitudes to religion. Describing the Western moral relationship with nature as *one of the most extraordinary developments in recent intellectual history,*[231] Nash documents a pre-ethical past where one's primary concerns were expressed through family, tribe, and region. This situation evolved into current ethical values relating to nation, race, humans, and to an extent animals. It may even be leading us towards a universal environmental ethic. These steps are as significant as the revaluation of slaves from owned chattels to humans with equal rights to all others. In this way, the concept of legal rights can be presented as a complement to the

---

[231] Nash (1989) Page 4

evolution of ethics and reveal sharp shifts in social perception often associated with 'movements' to convince, educate and lobby for change.

If *every great movement must experience the three stages of ridicule, discussion, and adoption,*[232] an ethical shift towards agricultural sustainability would seem to be barely out of the first stage, for unsustainable actions mock our well-meaning discussions. Doubts about technological salvation exist, but they are not within the output-oriented mainstream of agriculture. Nevertheless, we can see changes in Western ethical perceptions in the rejection of animal cruelty. From such consideration of the rights of living beings, we might next consider life-supporting matter and then ascribe rights to ecosystems, with persons speaking on their behalf. All of these matters affect agriculture and are related to the revolution that may make agriculture closer to the ideal of sustainability. So, rather than view agriculture as having sold its birthright for a mess of technological pottage, it makes more sense to view it as part of an evolving ethic of environmental rights - and indeed one that may yet end in a Jacob-Esau reconciliation.

Whether the liberal tradition of natural rights assigned to slaves, women and other groups can be expanded to non-human interests including all of nature depends on its validity within the evolving ethics of the West. While not using the language of rights, Greek and Roman philosophers assumed a pre-existing natural law distinct from the common law developed by states. This distinction allowed the Roman *jus animalium* to complement the *jus naturae* and *jus commune* with the effect that animals possessed independent natural rights. Following this ethic, it seems that legal courts in the Middle Ages heard cases of animals against humans.[233] However, the post-Reformation church assumption of the right to dominate

---

[232] Nash (1989) Page 8
[233] Stone (1974)

nature has reduced animal rights to the service of human needs. Such utilitarian values remained until after the seventeenth century when human rights became paramount.[234] Descartes was thus able to declare our superiority over all other life because of our ability for conscious thought.[235] So the West has moved away from the rights accorded to animals for nearly two millennia, and may now be slowly redressing this lapse.

As introduced in the preceding chapter, the continuity of pantheism was partly a reaction to the Renaissance developments, which in challenging anthropocentric perspectives have provided much of the basis for allocating rights to nature in the modern era.[236] In pantheist mode, Spinoza thoughts were amplified in Pope's popular lines:

Know Nature's children all divide her care;
The fur that warms a monarch, warm'd a bear.
While Man exclaims, 'See all things for my use!'
'See man for mine' replies a pamper'd goose:
And just as short of Reason he must fall,
Who thinks all made for one, not one for all. [237]

But questioning anthropocentrism has little impact on Western society when it clings to a belief in legitimized control of nature. Yet a change is occurring, influenced by biological logic that questions absolutist interpretations of Descartes and his followers. Western sensitivities seem to now allocate rights to higher or useful animals, particularly when they are as cute as dolphins. This is part of a continuum of gradual reassertion of animal rights through the eighteenth century when agriculturally useful animals were afforded an ethical status in a hierarchy that placed them above other life forms and below

[234] Passmore (1975)
[235] Descartes (1901) Page 170 ff
[236] Sessions (1977)
[237] Pope (1733)

slaves.[238] From this time, English laws that prohibited bearbaiting and cockfighting were initiated, although the issue of owner-inflicted cruelty to animals remained outside all such laws.

The Englishman Salt lived according to his principles of moral identification with animals.[239] His association of animal mistreatment with social decline may be seen as a precursor of deep-ecology.[240] Expressed as an enslavement of animal and even human rights to profit-motivations, his coining of the term 'liberation' allowed consideration of humans as oppressors who could make amends by accepting the oppressed as full members of the community. By modern times, *a century after Salt's 1885 rebellion against the English establishment, the idea of extending natural rights to include the rights of nature could no longer be brushed aside as a perversion of liberalism. For increasing numbers it was the new frontier of that philosophy.*[241]

Within agriculture, animal welfare became an accepted component of ethical management. While animal welfare is still not respected by all, it daily makes inroads into Western consciousness and in so doing increases our knowledge of ourselves. And ethical reassessment of animals stimulated consideration of all components of nature. This slow crawl towards an ideal of sustainable agriculture, while apparently secular in origin, honours our spiritual need to act in accord with nature.

So we see that secular philosophy is integral to an understanding of social and religious development as part of sustainability. The discussion may be summarized as follows:

---

[238] Bentham (1843)
[239] Salt (1921)
[240] Hendrick (1977)
[241] Nash (1989) Page 32

- If we conceive philosophy and science as expressions of spirituality in the same manner as religion, then we can see a gradual development of improved ethical approaches to each other and to nature.
- Within the evolution of ethics, just as we have conceived rights for slaves and animals, we may conceive the wider rights of nature and thereby find an ethical context for sustainable agriculture.
- An overemphasis on mechanistic explanations of nature has revealed that single-minded pursuit of profit from agriculture leads to unethical treatment of animals and nature that is not conducive to sustainable agriculture.

The emergence of rights is Western society's expression of an evolving individual self-consciousness. We should not expect to find the same reaction in cultures that have maintained greater balance between material and spiritual aspects of life. Yet today Western culture inevitably influences other cultures. In the ensuing cultural interaction, alternative means of the West conceiving man's integrity in nature became clear, as is discussed later. Meanwhile, the next few chapters amplify the evolving Western ethical understanding, beginning with the rise of secular ecological ethics.

Chapter 10

Sustainable Agriculture and Secular Environmentalism: Emerging Ecological Understanding

As scientists, many of us have had profound experiences of awe and reverence before the universe. We understand that what is regarded as sacred is more likely to be treated with care and respect. Our planetary home should be so regarded. Efforts to safeguard and cherish the environment need to be infused with a vision of the sacred. At the same time, a much wider and deeper understanding of science and technology is needed. If we do not understand the problem, it is unlikely we will be able to fix it. Thus, there is a vital role for both religion and science.

Leading Scientists[242]

Feelings of awe and respect may be shared by technologists and environmentalists, but perhaps their approaches to 'fixing' the environment may vary. Such statements as that above are useful dialogues between science and religion but ultimately assume that the two fields remain separate. Coupled with this non-unifying approach is a somewhat patronizing assumption among some parts of the environmental movements that social modification is part of the 'answer'. Thus the ideology of environmentalism may be traced from the types of rights - often guaranteed by constitutions - that are limited to white and propertied males.[243] Against this background, the speed of acceptance of rights in nature beyond utilitarian aspects is surprising,

[242] Open Letter (2000)
[243] Nash (1989) Page 33ff

considering the pioneering encounters of the new world with their inexhaustible natural resources by an anthropocentric culture. Perhaps it is this orientation to our own species that has favoured the first steps toward rights for nature in such forms as the establishment of national parks for recreational benefit. But the philosophers concerned are not necessarily those most popularized.

Folkloric comment often designates Thoreau as the source of modern Western sensitivities to nature, although his 'theological ecology' of a universe integrated under God's care derived much from Emerson, and his reputation is the product of Salt's early retirement writings.[244] Rather than project the next stage of rising Western social awareness onto Thoreau, we may do better to follow more worldly thinkers who influenced ethics within the continuing Western anthropocentrism,[245] for it is these that lead more clearly to the current contorted definitions of sustainable agriculture.

Our modern sensitivities to ecology are evident in the 1860s' works of Muir who proclaimed that all matter and life forms possesses a non-utilitarian value.[246] In so doing he indicated interdependencies between all matter within an ecosystem. His astute influence on environmental legislation was guided by an understanding of the spirituality of nature, which he described in his writing. His acceptance of small environmental legislation gains in the face of unbridled capitalism foreshadowed the techniques of present-day ecologists. The intellectual support provided by Darwin's contemporary work, which placed humans within rather than above nature,[247] provided a new context for Muir's arguments against hitherto unchallenged anthropocentrism.

---

[244] Torrey (1906)
[245] Marsh (1958)
[246] Fox (1985)
[247] Darwin (1874)

Our expanding view of life forms has remained anthropocentric, as is evident in the preferential rights ascribed to selected large animals. Even when the American Society for the Prevention of Cruelty to Animals was eventually established in 1866, it mixed its ethical debate with consideration of animal souls and their coexistence with humans in heaven.[248] And the ethical principle was oriented to human needs rather than any inherent natural rights of animals. From such a milieu, animals and nature ethics were reinforced as utilitarian in the form of economic evaluations of environmental conservation.

Economic evaluations are also applied to measures that purport to make agriculture more sustainable. Occasionally we consider the immediate durability of the agricultural ecosystem itself. But actions aimed at sustainable agriculture are not current with society's ethics and this will be of concern to democratic countries where the urban populace can now influence policies that affect agriculture. In terms of our discussion, it suggests that we may learn more about sustainable agriculture from developments of ecological thought beyond agriculture. By this I do not mean that we should slip into the terminology of 'ecologically sustainable agriculture', which is often a fiction of naïve political-correctness.[249] Rather I suggest that we consider general ecological ethics and their application to sustainable agriculture.

As is widely repeated, the word 'ecology' is derived from the Greek *oikos* meaning house, as is 'economics', and was probably first coined by Haeckel in 1866.[250] By the 1890s, its meaning consolidated around the interaction of organisms with each other and their environment. It was implicit in the means by which God managed creation according to the

---

[248] Steele (1942)
[249] Fitzpatrick (2000)
[250] Haeckel (1866)

theological ecology of Thoreau, and possibly even in the underpinning assumptions of Darwin. It finally entered the agricultural sciences through Bailey's influence in the 1920s.[251]

By 1927, Elton[252] explained ecology in terms of the 'food chain' which, notwithstanding some implications of a hierarchy of life forms, confirmed our mundane role within nature. The term 'ecosystem' was widely used by the 1930s to describe interdependent relationships between living organisms and non-living materials.[253] Whitehead[254] provided a philosophical basis in his description of the continuous flux of all things at all times, which defined the purpose of each object to be fulfilment of its relationship to all others. His hope that this organic approach to science would lead to recognition of the *intrinsic worth of every component of the environment* has yet to be realized in the sciences.

Around the same period, Schweitzer argued that our reverence for life should engender a responsible understanding of our influence on nature. Concerning our primary essential interaction with nature he noted that *the farmer who has mowed down a thousand flowers in his meadow in order to feed his cows must be careful on his way home not to strike the head off a single flower by the side of the road in idle amusement, for he thereby infringes the law of life without being under the pressure of necessity.*[255] The element of necessity accords with Eastern conceptions of not producing or acquiring above essential needs, and has spurred further thought around the rights of life forms to their place in an ecological cycle. Carson's influential book *Silent Spring,*[256] which warned of

[251] Bailey (1915)
[252] Elton (1927)
[253] Nash (1989) Page 59
[254] Whitehead (1920)
[255] Schweitzer (1950) Pages 189-190
[256] Carson (1982)

many environmental impositions related to agriculture, was dedicated to Schweitzer.

Even more influential, the works of Leopold re-introduced the agricultural sciences to an organic conception of nature with other species able to be considered essential to a well functioning ecology. He argued that land ownership should be abolished and provided a foundation for both a land ethic and a scientific rationale for an expanded environmental ethic in which the individual only exists as *a member of a community of interdependent parts*.[257] The poor or short-term management of privately owned land as a consumable commodity contrasted with communal associations of love and respect for land. This thought had hitherto been lost in mainstream Western agricultural writings.

As a biologist, Leopold understood the impossibility of avoiding human impact on the environment, and it was such practicality that fostered a wider understanding of the environmental ethic. However, his proposals were considered extreme in the post-depression 1940s of his USA. So mainstream ecology progressed through an alternative quantitative and reductionist approach, and adopted the methodologies of technological science. Agriculture adopted some of the approaches of ecology, such as the modelling of crop production systems, but largely omitted the evolving ethical aspects. Interdependence was thus reduced to a task of modelling in service to cellular and molecular research in a combined quest for higher productivity.

Post-war scientific and technological expansion further separated such applied ecology from its ethical associations, which in turn sought refuge in the humanities, religions, and the alternative organizations emerging around environmental issues. In an age of increasing economic rationalism, Krutch[258]

---

[257] Leopold (1949)
[258] Krutch (1978)

popularised the ideas of Leopold among philosophically oriented scientists and thus retained some linkage between the ethics and science of ecology. Subsequent thought recognized disease as part of a system rather than an enemy to be eradicated on all occasions, and heralded an 'enlightened anthropocentrism', a term which might also be applied to the works of Carson, in which she ultimately advocates a balance of nature which favours humans while minimizing the use of toxic chemicals.

However, behind *Silent Spring*, Carson's ethics followed those of Schweitzer, and her life and other writing bring the rights of nature into sharper focus, particularly where they confronted agriculture. Agribusiness opposed Carson's protection of insects and in their pique unwittingly amplified her voice to the public.

But it is always the utilitarian viewpoint that wins, which may be illustrated by Wilson's[259] successful argument that loss of species may work against future human interests as they might have potential for food, medicines and other purposes. Such selfish reasoning is claimed to be the only practical means by which ethical action can be stimulated. As scientific disciplines broaden their philosophical orientation over time, individual species may eventually be ascribed an 'existence value' simply on the basis of their presence in an ecosystem. Favoured by those who advocate sustainability, the sophistication of recognising the naturalness of the demise of some species can easily escape enthusiasts. The gradual appreciation in the broader public has reinforced these successive ethical shifts, which while apparently secular in their orientation, have often assumed a Christian cosmology.

This brief consideration of environmental ethics, means that:

---

259 Wilson (1984)

- Anthropocentric attitudes to sustainable agriculture have produced a utilitarian basis for valuing nature.
- Ecological conceptions of the large-scale environmental interventions of agriculture that assume the necessity to sustain original ecosystems are impractical when they ignore the necessity of food production.
- Manipulation of ecological rhetoric in agriculture ultimately exposes claims of sustainable agriculture to a public with an emerging ecological empathy and a poorly informed understanding of food production.

The artificial separation between secular and religious evolution that we have accepted here serves to highlight the role of philosophical interpretation in the fields that surround agricultural sustainability. The next two chapters continue this approach by looking at the interface of religion and philosophy with nature.

Chapter 11

The Religion of Sustainable Agriculture: Philosophy and Ethics

This is my simple religion. There is no need for temples; no need for complicated philosophy. Our own brain, our own heart is our temple; the philosophy is kindness.

Dalai Lama

To see the origins of our fascination with sustainable agriculture we must peer peripherally around our secular spectacles. As we have seen in chapters 3 through 6, a Christian environmental ethic is variously claimed as the source of modern Western concern for the environment. Within those claims, agricultural sustainability is also seen as a continuing Christian ethic – but as the more recent chapters indicate, it may more reasonably be described as building on ethical shifts in secular thinking. Even the religious tone of Leopold's observations[260] may be seen as a secular statement expressed in the language of his social group, for his main concern was the economic trivialization of conservation policies. In this chapter some of the religious language surrounding sustainable agriculture in secular contexts is discussed further in an attempt to understand its position in the modern emergence of ethical awareness.

If religion has determined most of the West's moral codes including rights,[261] it has also created an ethical separation between human rights and those of the environment. We have seen that revisionist interpretations

[260] Leopold (1949)
[261] Wood (1985)

now perceive the Judeo-Christian worldview as an extreme of anthropocentricity.[262] Following this reasoning, White sought a biblical basis for environmental care, and concluded that just as the Bible had been misused to defend slavery, so it had been erroneously interpreted to support human domination of nature. His thesis is supported by linguistic analysis of the operative Hebrew verbs of Genesis 1:28 where *kabash* is translated as subdue and *radah* as rule, whereas other uses of these words are more aggressive and are associated with violence and crushing an enemy, sometimes in association with enslavement.[263] Agriculture was a battle against nature. This allowed any manipulation of nature to be justified if it served man's ends. All that has changed recently is the realizations that this attitude can work against our long-term interests.

Respect for all beings and even inanimate objects as God's creation would once have been deemed animist by the Christianity. Now, as it responds to social pressures for improved environmental responsibility, the church seems set to embrace aspects of pagan religions that it once sought to eliminate. Nevertheless, the church has yet to catch up with secular thought and action. Past intransigence of Christianity in this instance has led many to seek new approaches through such means as:

- an interest in Eastern religions which have apparently retained an integrity between humans and nature,
- an interest in the animistic traditions of American Indians, and
- a reinterpretation of Jewish and Christian beliefs in positive environmental terms.

All three paths are contributing to environmental awareness in modern Christianity. White's search for a basis to

[262] White (1967)
[263] Humphreys (1971)

reform Christianity drew on these sources and assumed that, as environmental degradation appeared to be caused by religion, the solution must be similarly found in religion.[264] This assumption has allowed such influential changes as St Francis of Assisi being recognized, in 1980, as the Patron Saint of Ecologists - his mythical disinterested love of all life amplified the prudential ethic expressed in such general terms as the golden rule. However, the major reform of Judeo-Christian interpretations was an enhanced concept of stewardship based on Biblical reinterpretation.

Reinterpretation of Genesis 1:28 and related verses to favour trusteeship of nature rather than selfish exploitation retained the superior role of humans - in a new form as God's stewards. This was justified by the Genesis 2:15 command for man to *till and keep* the Garden of Eden. It is this association of stewardship with agriculture that has provided a significant basis for subsequent thought concerning sustainability. In this approach, the natural environment is sacred, as in pantheism. Of course, actions reveal the underlying motivation of sustaining profits, when required by an ethic of 'enlightened self-interest'.

It was the approach of enlightened self-interest that stimulated Dubos[265] to suggest the sixth century St Benedict of Nursia was a more ecologically representative patron than St Francis, insofar as he promoted the reverential draining of swamps, clearing of forests, and improvement of agricultural fields for human benefit - all actions that simple critiques of modern agriculture list as suspect.

However, before environmental saints were popular, the soil conservationist Lowdermilk closed the loop between sustainable agriculture and popular Christian thought of the 1930s when he argued that an omniscient God would have

---

[264] White (1967) Page 57

[265] Dubos (1972)

foreseen the impact of thoughtless agricultural management and have intended an eleventh commandment, such as:

> *Thou shalt inherit the holy earth as a faithful steward, conserving its resources and productivity from generation to generation. Thou shalt safeguard thy fields from soil erosion, thy living waters from drying up, thy forests from desolation, and protect the hills from overgrazing by thy herds, that thy descendants may have abundance forever. If any shall fail in this stewardship of the land, thy fruitful fields shall become sterile stony ground and wasting gullies, and thy descendants shall decrease and live in poverty or perish from off the face of the earth.*[266]

Strengthening of the link between conservation and morality caused some American churches to reconsider the implications of stewardship in the dim light of the great dust storms of the mid-west through the 1930s. By the 1960s, religion seemed to have embraced conservation, often using agricultural examples, and by the 1970s and 1980s the human-environment relationship was a major preoccupation of theologians, and to an extent, even church goers.

Coincident with Lowdermilk, parts of Christian theology morphed into a unity of God, humans and nature that ascribed religious status to enjoyment of the earth rather than its simple use.[267] This theological ecology considered all creation as of God and therefore worthy of reverence, which led directly to land misuse becoming a theological concern.[268] Old views seemed to have evolved into acknowledgment that even if humans are superior to the land and other beings, they remain inferior to God and are therefore charged with caring for his creation, for *God saw everything that he had made, and behold, it was very good.*[269]

---

[266] Helms (1984)
[267] Sittler (1954)
[268] Baer (1966)
[269] Genesis 1:31

Such universalism within eco-theology rendered destruction of the 'web of life' a sin. Agriculture received special dispensation as it was seen as essential to human life, although not essentially tied to private ownership of land. Theological thought then evolved to link environmental degradation with human suffering as a form of divine retribution. The new Christian theology of the 1967 conference *Christians and the Good Earth*[270] walked a fine intermediate line where the earth and its resources were seen as God's gifts, which humans are entrusted to manage. Once again, this is a version of stewardship. Later, the liberation of nature was paralleled with the past slave liberation rhetoric producing such terms as 'eco-justice', which in some conceptions cast agriculture as alienating the rights of nature.

As 'the inalienable rights' of nature pervaded at least USA religious consideration of the environment,[271] secular philosophies such as those of Whitehead re-entered Christian responses to the ecological 'crisis' from the 1970s. Natural rights liberalism emerged to conceive that *the non-human world has just as much right to its internal integrity as does the human world [and] human beings transgress their divine authority when they destroy or fundamentally alter the rocks, the trees, the air, the water, the soil, the animals – just as they do when they murder other human beings.*[272] With such religious trappings, an eco-theology developed for some Christians, who also saw agriculture as a holy interaction with the sacred resource of farmland. In turn, this reinforced the stewardship ethic of sustainable agriculture in the mid-west USA.

Through the same period, increasing appreciation of wilderness areas evoked Native American beliefs. Wilderness itself was defined as a church and requiring protection.[273]

---

270 Stefferud (1968)
271 Santmire (1985)
272 Brockway (1973)
273 Hughes and Swan (1986)

Asian religions began to influence modern Christian thought and encouraged the faithful to see themselves as a component of nature, thereby further challenging notions of rights over nature. Such doctrines as *ahimsa* (universal non-violence) widened consideration of all living and life-supporting components of an environment, as is considered further in a later chapter.

Familiarity with Asian traditions of such popular Christian thinkers as Thoreau and Schweitzer provided a foundation for understanding such exotic figures as Suzuki. Suzuki's influence through the 1950s inspired the reactionary rhetoric of a generation which forsook churches for the open air, expounded nature worship, and forged the composite religious viewpoints based on Buddhist, Native American and inherent Christian ideologies that remain evident in the works of Snyder.[274]

Whether this was pantheism or stewardship matters little. More significant were common beliefs in sound land management and an abhorrence of land ownership, animal enslavement and financial valuations of habitat destruction. In each case, sustainable agriculture was conceived as more than simply maintaining profits from agriculture. Environmental awareness as a context for sustainable agriculture appears earlier in the USA than in most other Western countries. This produced an informal philosophy developed outside the mainstream which provided a basis for environmental ethics. It also informed debate in cultures with strong natural rights traditions, such as the United Kingdom, Canada, Norway, the USA and Australia.

Animal and ecological rights then extended to universal conceptions of life and culminated according to some, in the often misunderstood Gaia hypothesis. Initially limited to humans and animals, the argument required a being

---

[274] Snyder (1974)

to have interests and be a beneficiary in order to have rights; plants were excluded for their cognitive inadequacies, yet protected according to human interests. This thinly disguised reassertion of human domination echoed Descartes' earlier exception of animals from the moral community using the same cognitive argument. Nevertheless, legal challenges in California confirmed that rights only applied to human injury – injury to nature alone did not constitute an infringement of any extant right. In the mid-1970s humans were re-construed as agents to represent the interests of inarticulate natural forms such as ecosystems and trees. The argument gained advocates through associations with higher animals, in particular dolphins, and led to a 1979 proposal that the USA Constitution be modified to provide for all wildlife having the right to a natural life within an ecosystem.[275]

The church was at best a follower, as clearly seen in the manner in which animal rights captured popular thought and action. In his strenuously secular manner, Singer argued for the liberation of animals on moral grounds, producing a religious-style response across much of society reminiscent of liberation movements for American blacks and women.[276] Mistreating humans, laboratory animals or agricultural livestock were seen as part of an ethical continuum.[277] There was no need to fall back on old arguments of the harmful effects of animal cruelty on humans.

Nevertheless, much effort was wasted in defining sentience as a means to ascribe rights to individual species. Similarly, confusion over metaphors for environmental care fostered feminist interpretations of environmental domination. 'Rape' of virgin land and abuse of women in Western society were contrasted with supposedly different traits in Asian

---

[275] Nash (1989)
[276] Singer (1975)
[277] Harrison (1964)

cultures[278] as part of an eco-feminism that sought a home in the deepest of deep-ecology notwithstanding its quite different intent.[279]

Deep-ecology accommodated humans within the ecosystem and included necessary killing and exploitation as a function of human survival. Killing was ethically acceptable where it was essential to basic needs, producing the conclusion that much human environmental impact was acceptable because it was inevitable. Environmental ethics therefore became the means for restraining excessive human exploitation of nature. However, the more popular interpretation became one of protecting nature so that it might take care of us. National parks were therefore justified on their recreational capacity. By this stage, Westerners could 'believe' in a deep-ecology ethic without the need for institutionalized religion. Seeing oneself as part of nature made stewardship an act of self-interest.

A pragmatic conservation movement had now emerged with an enlightened self-interest approach for political advocacy that required compromise of deep ecological principles.[280] By the late 1970s, animals had rights similar to 'legal incompetents' that required human guardianship, although philosophical entertainment continued in the consideration of equality of rights for human pathogens.[281] As the only moral agent in the biosphere, humans assumed responsibility for ethical oversight. This tautological conflict of interest contextualizes much of the discussion about rights around manipulation of the environment to improve the human lot.[282]

---

[278] Marietta (1984)
[279] Naess (1973)
[280] Capra (1975)
[281] Callicott (1988)
[282] Passmore (1974)

Religious-like rights to *life, liberty and the pursuit of their own style of happiness* raised the integrity of the ecosystem to a level above the rights of any individual life form – and went beyond the limits of popular moral tolerance.[283] Pragmatic day-to-day business could ignore environmental arguments that lacked a public voice, although the direction of change had already been established. In terms of agriculture, the essential need for food allowed it to be treated as a special case, and in any case, its major impacts on the environment had long become irreversible.

The special-case approach to agriculture means that environmental issues are usually considered as singular problems, an approach not conducive to sustainable agriculture. At the same time, philosophical consideration of agricultural productionism, land and nature stewardship, and holistic perspectives provide arguments but no real guidelines for ethical action. And the result is a reliance on bureaucratic interpretations of *regulatory policy for agricultural biotechnology [that is] buried deep in the forest of government acronyms.*[284] Perhaps scepticism is understandable when we recall that Leopold and Schumacher exaggerated the attitudes of producers, researchers, and academic environmental ethicists in order to make their points. The technique of argument has not been missed by agricultural advocates who now use the language of ecology and social justice to justify continued expansion of production.

Continuing high levels of agricultural production requires continuous funding of agricultural research, which is commonly argued on the basis of saving land for nature. In other words, if we can produce more grain from an area, we do not need to expand into forests and other sensitive environments.[285] The argument builds on Locke's philosophy

---

[283] Nash (1989)
[284] Thompson (1995)
[285] CGIAR (1997)

– [he] that encloses land and has a greater plenty of the conveniences of life from ten acres than he will have from 100 left to nature, may truly be said to give 90 acres to mankind: for his labour now supplies him with provisions out of ten acres, which were but a product of an hundred lying in common.[286] This is not simply sustaining 90 percent of nature and sacrificing 10 percent, but is an early acknowledgement of rising population and the Protestant work ethic linked to God's grace – it was this very industriousness that produced industrial agriculture.

Intensification of agriculture has produced such efficiencies as the commonly claimed success of feeding a nation from the work of only four per cent of its population.[287] But that four per cent is a fraction of the food supply network. Others include buyers at various levels, retail outlets, processing facilities, packaging, transporters, infrastructure supporting all of these and other fields from physical structures through to banking and input suppliers including chemicals, feeds and pharmaceuticals. Within this paradigm, stewardship, rights and integrity with nature cannot be credible concepts unless they include all involved in the food industries, including consumers. We are far from such holism in modern agriculture.

An holistic view of agriculture sees it as both a component of a larger *biotic community and as an ecosystem in itself.*[288] Ethicists following the ecosystem conception challenge traditional Newtonian approaches to understanding parts of natural processes and manipulating them for agricultural gain.[289] Systems theory is often interpreted as a reaction against Newtonian reductionist approaches, and in fact has long been used in those parts of agricultural education that have not been captured by reduction-based specialists.

---

[286] Locke (1690)
[287] Berry (1981)
[288] Thompson (1995)
[289] Callicott (1990)

Systems agriculture may be a meeting ground between the enlightened self-interested views of stewardship and holistic views of ecosystems. However, in production-dominant situations, we know that *philosophically this use of the word 'sustainability' makes no conceptual advance over utilitarianism.*[290] While external observers may comment on the actions that a farmer might take, his own objectives in fact determine his actions and are necessarily influenced by his knowledge, personal requirements and even whims. The ethics of seeking to maintain stability under the guise of sustainability are seriously undermined by their separation from the subject they seek to influence.

A sustainable system is normally defined within specific limits in the agricultural sciences. However, biological systems do not have specific borders. We can probably assume that while inputs are maintained, a reliable prediction of output levels can be made, but we never know all of a system's inputs. We could also assume that the various artificial interventions of agriculture operate as a stable ecology with the system's homeostatic mechanisms. But it is naïve to assume that agricultural systems are closed, controllable, or predictable. It appears that current ethics of agricultural sustainability are based on such flaws. Limited knowledge, let alone wisdom, poses an ethical irony.

There is a further irony in the observation that we may learn about sustainable agriculture from societies now considered unethical, because of their reliance on strict social hierarchies for example. Where agriculture was practiced by smallholders who worked close to their land and ate their own produce, it seems that agricultural systems were relatively durable. Where an external demand for increased production exceeded social or ecological tolerance, agricultural production

---

[290] Thompson (1995)

ultimately failed.[291] Agricultural history suggests that wherever output becomes an overriding objective, natural systems eventually change substantially. This is why a *system-describing sense of sustainability must be given an implicit priority over the goal-describing sense.*[292]

For this systems-describing sense of sustainability to have application, comprehensive models are needed that include technical and biological parameters as well as human values, ethics, and other philosophical viewpoints that may impact on the sustainability of the agricultural system. When conceived from this iterative perspective, *the feedback mechanism that makes modern life responsive to environmental threats*[293] may be conceived as part of the system. If this is true, the modern ethical stance might seem to have insinuated sustainability into the notion of progress. But we do not even come close to creating models that include all variables, not the least because we understand very little of the processes of the systems themselves. For the time being, agricultural science seems destined to continue its perpetual attendance on each new problem as it arises - problems often caused by preceding 'solutions'.

Today's agricultural science is problem-solving technology. It worries little of philosophy or ethics beyond determining means of continuing its science in the face of public constraints. In this way, it does not differ from other technological human endeavours. To discuss agricultural sustainability in that context defines it as 'a potential but indefinitely deferred end point' rather than a specifically defined activity. Perhaps the benefit of the debate is ultimately raising awareness and heightening environmental responsibility. The issues canvassed in this chapter may be summarized in the following points:

---

[291] Falvey (2000)
[292] Thompson (1995)
[293] Thompson (1995)

- It is difficult to separate religious from secular influences on eco-theology which conceives agriculture as an essential activity that only becomes sinful if it unduly disrupts the 'web of life'.
- To consider nature sacred was criticized as pantheism by the church, which re-emphasised an ethic of stewardship that was rapidly overtaken by public and philosophical conceptions of rights for all elements of nature.
- Consideration of ethics specific to agriculture has produced little of consequence as it has largely ignored the motivation behind actions, even those claimed as sustainable.

With the role of religion now assumed by a philosophy that interprets societal feelings, a wider debate about the 'liberation' of nature provides another context within which agriculture operates, as is discussed in the following chapter.

Chapter 12

Liberating Nature: Our Rising Awareness

Reflective virtue is simply an acquired clarity of the intellect, and moral virtue is constant warmth of heart kindled by that clarity. We should remember that of the human virtues none is more precious than discrimination … For everything is an obstacle and nothing of use to a man who cannot distinguish the good from the bad and separate the bad from the good.

Marsilio Ficino

Whether seen as religious or secular, the liberation arguments that found expression in the USA fight against slavery offer us another perspective on the West's changing perceptions of the environment. It seems that once it is generally felt that a natural right has been denied, moral outrage effects a revision of oppressive rules. Today the rights of land and livestock conflict with Western concepts of property in the same way that equal rights once conflicted with slaves being considered as property. But solutions to conflict are first approached technically rather than ethically. And this technological orientation since the 1960s coupled with faddish approaches to ecology caused Bookchin[294] to observe that, contrary to well-meaning policies, dominance of nature had accelerated.[295]

In the wider world, the green politics of West Germany linked peace, anti-nuclear policy, women's rights and

---

[294] Bookchin (1986)
[295] Bookchin (1988)

environmental ethics. This encouraged similar actions in Australia, the USA and elsewhere. As popular concern grew, so scientific investigation of the rate of species disappearance, for example, accelerated and in turn provided a basis for legislative action to protect the integrity of an ecosystem. However, in the final analysis, selfish ends were necessary to win such gains and thus the principal argument against loss of a species was the loss of some potential value to humans. For example, the USA Endangered Species Act of 1973 aimed to preserve species of 'aesthetic, ecological, educational, historical, recreational and scientific value to the nation and its people'.[296] Nevertheless, the legislation covered all land including private lands, which was revolutionary. Too revolutionary for its time and the legislation was soon diluted by vested interests. Piqued environmentalists used this reversal as a justification for protests and physical intervention through such means as Greenpeace and militant animal liberationists.

Animal liberation movements grew through the 1980s. A common argument was that animals had rights to liberty and fulfilment and it was therefore inappropriate for them to be held against their will, except for essential purposes. Captured dolphins were thus freed and youth were motivated to 'liberate' animals using the rhetoric of slavery abolitionists. Agricultural industries, from chicken to dairy farms and abattoirs were targeted as immoral.

At the same time however, the general environmental debate developed in a more sophisticated manner than did considerations of agricultural sustainability. The environmental movement of the late 1980s invoked the threat of nuclear war to all of nature, not only humans. Thus the themes of peace and environmental responsibility were united. The coincident rising of Western awareness of Eastern

---

[296] Nash (1989) Page 176

philosophy then allowed parallels to be drawn between animal liberation and Jain and Buddhist understandings of *ahimsa* (non-violence). Buddhist conceptions of interdependence were also interpreted as ecology and a form of wishful reasoning emerged which concluded that technological prowess produces environmental downfall. Such an emotive conclusion that opposed mainstream Western agriculture was easily ignored. Nevertheless, animal and nature rights continued to attract ever wider support.

The trend to accept rights accorded with the observation that , *if the abolition of slavery marked the limits of American liberalism in the mid-nineteenth century, perhaps biocentrism and environmental ethics are at the cutting edge of liberal thought in the late twentieth.* Just as American slaves had no natural or other rights including loss of life, except tenuously as valued property, so in agriculture compensation for livestock loss through inadvertent actions of shooting, dangerous driving or irresponsible use of poisons ascribes the only right as value to the owner. The animal liberationist Ryder observed in 1975 that *when we examine the arguments used by slave-owners in the past, we see a striking similarity with the view expressed today by those who defend the exploitation of animals in factory-farms, the fur trade, [and] laboratories ... above all else, it would be stressed [that either human or animal] slavery was necessary for economic survival.*[297] Humane treatment of slaves began as good business to ensure a healthy workforce, a form of economic stewardship. Modern agriculture is now at the same stage of considering animal welfare as good business, for a less stressed animal is more productive. In this way, animal science and industry insinuate themselves into sustainability rhetoric and claim responsiveness to changing social mores.

We might thus find the social mores of slavery and its abolition instructive for our current purposes. If we view the

---

[297] Nash (1989) Pages 200, 202

American civil war as one of a conflict of ethical ideas rather than preservation of the union, the civil war is linked to the violence associated with ethical change in Western society from the Magna Carta onward. The apparent low probability of widespread violence in defence of animal or ecosystem rights might then be parallel with the apparent unlikelihood of civil war in the early days of abolitionist activity. Modern 'liberationists' argue that it is in our own interests to be considerate of animal and environmental rights. Again, this is the same as early arguments for slave welfare. But if we consider Kant's reasoning that moral law is absolute and followed as a natural process without conflicting motivations including self-interest, then we realize that any law which does not engender its own observance is not really a moral law. This is important to our quest for sustainability based on self-interest. Yet we seem to assume that self-interest has been a sound basis for ethical progress through history.

Most historical considerations of Judeo-Christian influence on Western environment ethics centre on the stories of Genesis, as discussed in earlier chapters. In these stories, God grants humans dominion over the earth and its inhabitants, which is emphasized in the proclamation that *fear of you and the dread of you shall be upon every beast of the earth, and upon every fowl of the air, upon all that moves upon the earth, and upon all the fishes of the sea; into your hand they are delivered.*[298] As if this was not clear enough for Western culture, Aquinas skilfully melded Christian and Aristotelian thought to conclude that *since nature makes nothing purposeless or in vain, it is undeniably true that she has made all animals for the sake of man.*[299] It is a powerful argument that licenses domination and contrasts with the caring alternatives of Saints Francis and Basil who advocated unification with nature – neither of

[298] Genesis 9:2
[299] Aristotle (1916) Page 16

which have had much influence on our relationships with other beings and nature.[300] Of course one can find caring attitudes in the same ancient books where the Sabbath allows rest for working animals.[301] Similarly *Deuteronomy 22:10 forbids the farmer to plough with an ox and a donkey yoked together because the practice would obviously impose great hardship upon the weaker animal.*[302] But these are utilitarian or if you like, self-interested, explanations.

The utilitarian approach undervalues nature, from agriculture, to gardens, to wilderness areas when it only considers their aesthetic or commercial value. Agriculture is undervalued when land is only valued for its productive capacity. Of course, it can be argued that the assigning of value is the economic process at work, and that society is always revaluing everything. In Aristotle's time, differential values were assigned to individuals according to whether they were slaves or barbarians, who it was reasoned, did not share the same feelings as elite Greek citizens. Most people do not think that way today. As views of the inferiority of animals change in response to 'animal liberation' arguments[303] a revaluation of the feelings of agricultural animals may well arise.

Western culture has long assumed that other beings feel pain less than we do. Insightful and sensitive persons across the ages have known better. Now in the face of evidence, the 'rights' of animals are understood to mean that they should not be subjected to needless pain. Eastern traditions obviously incorporate a different ethic concerning animals, sometimes invoking the concept of sentience. But Western interpretations of these teachings tend to follow the same hierarchy of life forms with humans at the top followed

---

[300] Singer (2000)
[301] Exodus 23:12, Deuteronomy 5:13-14
[302] Gordis (1986)
[303] Singer (2001)

by species that humans prefer. And such an understanding is difficult to extend beyond living beings to other elements of nature. Separation of life forms into sentient and non-sentient or animate and inanimate does not seem to lead to any practical or rational understanding of living in harmony with nature. It easily leads to the argument that one should *save the whale but kill the mosquito* at the same time as arguing to protect biodiversity.

The intrinsic value of diversity and complexity is in the whole, and when it is fragmented it into individual parts that value is destroyed.[304] If humans are part of that whole, then it follows that all beings should be treated as fellows, be they animals or other beings. They have a right to exist. Recognizing that all beings affect others, one should strive to minimize unnecessary impacts. Easily said, such sentiments ignore humans' natural reticence to accept upsets to existing comfortable arrangements.

The common preference for the *status quo* is evident in the agricultural sphere in approaches to food shortages. The agricultural sciences continually seek to produce more, ostensibly to meet supposed shortfalls in supply – and that is about it. They do not address such issues as equitable access to food or land, political control of food to command power, or exploitation of farmers in less-developed countries to serve Western markets. To change any of these could affect the existing order of social systems, comfort, and wealth. Yet *preventing the starvation of millions of people outside our society must [ethically] be considered at least as pressing as upholding property norms within our society.*[305] Of course, while exposing a general hypocrisy, this is no worse than in previous generations.

---

[304] Singer (2000) Pages 86-102

[305] Singer (2000) Page 112

Religious philosophers from the time of Aquinas, notwithstanding his conclusion that nature existed for human benefit, have argued that we must share. *Now, according to the natural order instituted by divine providence, material goods are provided for the satisfaction of human needs. Therefore the division and appropriation of property, which proceeds from human law, must not hinder the satisfaction of man's necessity from such goods. Equally, whatever a man has in super-abundance is owed, of natural right, to the poor for their sustenance. So, Ambrosias says, and it is also to be found in the Decretum Gratiani: 'The bread which you withhold belongs to the hungry; the clothing you shut away to the naked; the money you bury in the ground is the redemption and freedom of the penniless'.*[306] Thus we return to the conclusion that taking in excess of needs is immoral; the reciprocal of this is charity - willingly giving from resources that we might think we own.

Relative generosity impressed the medieval Jewish moral philosopher Maimonides who categorised charity in a seven-stepped *Golden Ladder* rising from reluctance to detached giving. His seven steps were:

1. Reluctant giving.
2. Cheerful giving insufficient to the need of the distressed person.
3. Cheerful and proportionate giving, but only when asked.
4. Cheerful proportionate giving without being asked but giving in a manner that humiliates the recipient.
5. Giving to an unknown beneficiary who knows who is the benefactor.
6. Giving to a known beneficiary who does not know who is the benefactor.
7. Giving to an unknown beneficiary who does not know who is the benefactor.

[306] d'Entreves (1948) Page 171

An even higher weighting is placed on anticipation of need and acting to prevent the need for charity.[307] Similar analyses can be found in Christian works and Buddhist concepts of *dana* and insight.

Clearly, altruism is a high moral action that is difficult to explain from conventional evolutionary biology. *For the first time since life emerged from the primeval soup, there are beings who understand how they have come to be what they are. In a more distant future we can still barely glimpse, it may turn out to be the pre-requisite for a new kind of freedom: the freedom to shape our genes so that instead of living in societies constrained by our evolutionary origins, we can build the kind of society we judge best.*[308] Sustainable agriculture from that viewpoint might simply be a small step towards an improved society, as it is less selfish than short-term exploitation. But its usual selfish associations place it low on the ladder of charity.

For sustainable agriculture to be a step toward an improved society its actions must be moral. To act morally has more benefit than simply feeling good from an altruistic action – its effects further moral behaviour, in the manner of spiritual practices. Aristotle, Buddhism and Christianity all see virtue as created by its practice, which Aristotle likened to becoming accomplished at playing the lyre by practising it.[309] Though mired in self-interest, perhaps our liberation of animals and nature may be a first step towards such a virtue. So, we might summarize our discussion as follows:

- Modern secular philosophy interprets moral trends and raises awareness of the rights of animals and nature, at least in utilitarian valuations.
- Animal agriculture that neglects the rights of animals assumes that animals do not feel pain in the same manner

---

307 Twersky (1972) Page 136
308 Singer (2001) Page 282
309 Singer (2001) Page 262

that we do, as was once argued about slaves, and thereby infringes the West's emerging moral code.

- Sharing and other deliberately generous behaviour improves natural morality, and in this regard, sustainable agriculture might be seen as a step towards wider interaction with nature.

Rights and liberation of nature are just two means of addressing the issue of sustainability, and are inherently flawed by the fact that is ultimately we who decide on all other rights in nature. Other secular philosophical influences on agriculture include international economic paradigms. We might therefore expect approaches to economic and sustainable development to inform us further on sustainability. This is considered in the following chapter.

Chapter 13

Sustainable Development: Having it All?

Sustainability ... means not taking from the earth, from the world, from society, from each other, from life more than we give back. But when industrial society uses the word, it means the sustaining of itself, no matter what the cost. It means sustaining privilege, sustaining poverty, sustaining abuse of the earth, sustaining inequality, sustaining starvation, sustaining violence. To sustain the existing system ... is neither conservative nor sustainable.

Sivaraksa[310]

The Western model for continued growth not only underpins the means by which lifestyles are maintained, but also means of fostering a sort of global utopia. This is integral to the Western worldview, even if some beneficiaries nominally dissent from it in favour of nature. And attitudes to nature fall within a spectrum of diverse responses to environmental discourse. A useful classification of environmental attitudes is presented below.[311]

	Reformist	**Radical**
Prosaic	Problem Solving	Survivalism
Imaginative	Sustainability	Green Radicalism

From such an analysis, we can understand more about our current infatuation with sustainability. The four possible

[310] Sivaraksa (1993) Pages 96-97
[311] Dryzek (1997) Page 123

outcomes from the model are; *problem solving* - which refers primarily to policy and market interventions, *survivalism* - the Club of Rome's (see next chapter) assumption of an exhaustible natural resource stock, *sustainability* - approaches without apocalyptic scenarios, and *green radicalism* - which rejects common conceptions of human relationships to the natural environment.

It is usual to ascribe the term 'sustainable development' to the Brundtland Report (see next chapter), although its use and certainly the concept long predated that report. By the early 1990s, the terms had proliferated to at least 40 different types, which neither the USA Academy of Sciences nor UNESCO was able to unify into a single definition. To arrive at the term, the Brundtland Report assume no absolute limits to growth, and considered current limits as imposed by technologies and social organizations that could be changed with ingenuity.[312] In this way, this influential report supported continuous progress within sustainable development.

Green Radicals saw the linking of development to sustainability as a compromise. Nevertheless, the rhetoric of Brundtland flowed into aid programs, international development agencies, and domestic policies in the OECD and later, other countries. Thus deep-ecology, cultural eco-feminism, bioregionalism as a sense of place, lifestyle, eco-theology, eco-communalism and much of the philosophical and theological outcomes described in previous chapters were downplayed as unnecessarily 'radical'. Yet these *green radicals* had their own definitions of sustainable development. And it is this diversity of definitions that has rendered the term so easily captured by various interests, leading for example, to business profitability and economic growth becoming institutional definitions of sustainability. Recognizing this

[312] Dryzek (1997) Page 30

confusion, Dryzek proposed that *ecological modernization* may be a more precise description than 'sustainable development'.

Ecological modernization is based on the precautionary principle that scientific uncertainty should not be used as a reason for inaction in protecting an environment. It acknowledges that market forces conflict with environmental protection and remediation, and that social and financial realities need to be balanced with modifications to environmentally unsound actions. Thus the interaction of consumption, production, resource depletion, and pollution is considered as a dynamic system within consumer- capitalist politics – a political system that itself may be anathema to many *green radicals*.

Extension of this *ecological modernization* under the banner of 'sustainable development' to less-developed countries through aid and international assistance naïvely assumed that the elements of Western governance systems and privileged market access exist in these countries. The result was often a disappointment to both economic development and environmental protection advocates. So far, the grail of a sustained technological utopia is far from being anything more than an unattainable aspiration. Nevertheless, 'sustainable development' continues to be a required 'research target' that is fully expected to be defined and achieved. Such utopian dreams pervade considerations of sustainability.

Utopian hopes for sustainability of a social hierarchy would do well to recall that the 'utopia' of King Asoka in ancient India was founded on violence before his change of heart. In any case, Asoka's new approach proved institutionally unsustainable within two generations. As Aldous Huxley highlighted in his novel *Island*, utopian states are ultimately and inevitably powerless against outside oppression.[313] The fictional militarily-aware utopia of seceding

[313] Huxley (1947)

states of the USA described in *Ecotopia*[314] more closely resembles modern views of sustainable development – including the utopian flaw.

As for utopian communities, sustainability is an ideal that is useful for decision-making but not for expectations. Plato appears to have recognized this in his advice to both Dionysius the Elder and the Younger to rule as philosopher-kings while remaining aware of human actions.[315] And our understanding of Plato's utopia and metaphysics, which so influenced Christian theology, is now considered to conflict with Plato's other writings that show this as a misconception of Socrates' insights.[316] So we might conclude that if utopia does not exist, neither does sustainability.

This chapter may be summarized in the following points:

- Sustainable development is instilled into the Western worldview.
- Limits to continued growth have been reconceived as temporary constraints of current technologies and social arrangements that will be overcome by constant innovation to sustain whatever is desired.
- In reality, no culture has ever been sustainable, and beneath the veil of definitions for sustainability, it can be seen as a pragmatic component of consumer capitalism that is being extended across the globe.

As the West extends its worldview to the world, it provides itself with a basis to examine its own attitudes. Does it seek to pragmatically sustain the 'natural' environment, or is it justifying its old ways in new words? The following chapter considers international words and actions.

---

[314] Callenbach (1975)
[315] Ryle (1966)
[316] Ellis (2001)

Chapter 14

Sustaining Our Role: Global Sustainable Development

That wisdom embodied in Nature is not one construed of theorems but is a unity, is not a wisdom compiled of diverse parts, but is rather a unity expressing itself in diversity.

Plotinus

International development is conducted through the multinational development banks, aid agencies, bi-lateral aid programs, regional agencies, and a broad industry of delivery mechanisms ranging from consulting groups to non-government organizations (NGOs). Its rhetoric to 'advise and assist' has been parodied as 'devise and insist' – thus exposing its pragmatic transplanting of Western values. One of these areas of insistence is now termed 'sustainability'. It builds on the logic of the Club of Rome's Malthusian update of the 1970s and the later Brundtland Report on international development.

The concerned and influential group that called itself 'The Club of Rome' published 'The Limits to Growth' in 1974. They predicted that the limit beyond which growth of human population could not be sustained with existing policies, lifestyles and political systems would arise in the twentieth century. It introduced a concept of *ecological and economic stability that is sustainable*, which could be approached if the *basic material needs of each person* were satisfied with *equal opportunity to realize … individual human potential*.[317]

---

[317] Meadows et al (1974) Page 24

The Club of Rome implied an equilibrium state being reached through worldwide birth control, an average of two children per family, and maintenance of industrial output at 1975 levels. Any excess production capacity over basic needs was to be applied to useful consumer goods. Industrial investment was to be linked to depreciation rates.[318] Notwithstanding its neglect of human ingenuity to forestall yet another Malthusian prophecy, the message was distorted into a problem that could be solved by technology on a project basis. The large agricultural development projects of the 1970s and 1980s were to forestall the apocalypse. In this way, the ecological and equity messages of the Club's call were downgraded to motherhood rhetoric in a consumerist lexicon. It was a popular deception, for such a solution *requires a change only in techniques of the natural sciences, demanding little or nothing in the way of change in human values or ideas of morality*.[319]

The next significant event was the World Commission on Development's study that has become more widely known as the 'Brundtland Report', and which was published as 'Our Common Future'. Its historical context has latterly been cast in terms of United Nations' initiatives, which probably reflects the co-incidence of diverse international concerns. In this report, the social concern that found expression in the Club of Rome's work was redirected to the environment. This in turn stimulated the preparation of policies related to 'sustainable development', a term that had been used from the late 1960s and that was used in documents that precipitated the 1972 Stockholm Conference on the Human Environment, the 1980 World Conservation Strategy, and the establishment of the United Nations Commission on Environment and Development in 1983 - which published the Brundtland Report.

---

[318] Meadows et al (1974) Pages 166, 170
[319] Hardin (1968)

The Brundtland Report defines sustainable development as *development that meets the needs of the present without compromising the ability of future generations to meet their own needs*. Thus it enshrined the concept of stewardship. It was a powerful compromise of pragmatism and idealism and soon assumed a mantra-like acceptance. A successor to the Report according to some observers was the United Nations Conference on Environment and Development, or the 'Earth Summit', held in Rio de Janeiro in 1992.

In support of meeting the essential needs of *food, clothing, shelter, jobs,* the Brundtland Report drafted an addition to human rights under the rubric of the environmental premise that *all human beings have the fundamental right to an environment adequate for their health and well-being*. In terms of sustainable agriculture, it espoused *policies to ensure that food is produced where it is needed and in a manner that sustains livelihoods of the rural poor.*[320] Well-meaning as such sentiments were, they were easily transmuted into domestic environmental concerns in more-developed countries, which precipitated the export of dirty industries to less-developed countries. But it also engendered a general rising of environmental awareness globally. Most significantly, the Brundtland Report revised absolute limits to growth to a faith in the development of technological and social solutions to environmental problems.[321]

The wider environmental sentiments of Brundtland found consensus in the Earth Charter proposed to the Rio Earth Summit in 1992. The Charter elicited ethical principles for a sustainable way of life that appealed to a cross-section of secular and religious groups. However, the Charter was not adopted at the 1992 Summit and a program continues through the Earth Charter Commission formed in 1997 to promote a

[320] WCED (1987) Pages 87, 392, 162
[321] Dryzek (1997)

worldwide dialogue on shared values and global ethics. The aim of that dialogue is a global people's treaty that promotes commitment to the Earth Charter in lifestyles, professional and organizational work ethics, educational curricula, religious teachings, public policy, and government practices.

The Earth Charter presents sustainable development as *full human development and ecological protection* underpinned by holistic thinking, *freedom, justice, participation, and peace as well as ... economic well-being.*[322] Inevitably, sustainable development was interpreted as a materialistic action. It would, among other outputs, *enhance individual and community well-being and welfare by following a path of economic development that safeguards the welfare of future generations, provides for equity within and between generations, and protects biological diversity and maintains essential ecological processes and life-support systems.*[323] The rhetoric entered the World Bank as *the proposition that economic growth, the alleviation of poverty, and sound environmental management are in many cases mutually consistent objectives.*[324]

If a principle of 'sustainability' is the maintenance of present Western lifestyles,[325] then we should expect reduction of poverty in less-developed countries through economic growth to have some limits. The limits to economic growth in less-developed countries appear to be *constrained to satisfy sustainability criteria.* But of course, by narrowing the initial principle in such a manner, the conclusion is pre-specified - growth will be according to Western pleasure. It is akin to definitions of poverty, which when limited to financial measures produce solutions expressed in financial terms. Thus an assumed requirement for economic growth has come to transcend consideration of other social aspirations and

---

[322] Earth Charter (2000)
[323] ATSE (2001)
[324] World Bank (2000)
[325] ATSE (2001)

maintains a narrow logic that insists on *sector-based sustainability codes and guidelines*. Such reduction of a holistic approach has caused some scientists to consider sustainable development to be an oxymoron - a statement that, in the continued spirit of misunderstanding, is taken to mean that most environmental change may already be irreversible. But the very real likelihood remains that sustainability may not be compatible with the growth-based scenario.

As an illustration of the pervasiveness of this attitude across the gap between rich and poor countries, we may at the example of Thailand, which has aligned its definitions of sustainability to economic progress. Thus sustainable development is *sustainable use* of natural resources with science and technology used as a tool to provide information about resources, their management and use. For example, *in order to contribute to sustainable social and economic development, the environmentally sound application of technology, including biotechnology in the conservation and sustainable use of tropical bioresources requires human resource development, capacity-building and creating greater awareness and understanding among the general public and key decision makers*.[326]

So the uncertainty of change that was once a primary focus for agricultural science to understand and work within becomes the enemy of sustainability. Here we may see the expanding and unnatural divide between 'nature' and human interventions. The example of global warming is apposite. Its science defines changes and causal relationships in a complex model that may assume a role in predicting future changes, but too easily shifts to *sustainability of the atmospheric environment* with a reduced regard for natural variations.[327] The attitude also pervades management of natural resources, which is determined to be *measuring the sustainability*[328] of

[326] Sabasri (1994)
[327] Zillman (2000)
[328] Reichelt (2000)

management actions, or *protecting [a] resource for continued long-term benefit*.[329] Thus human control of natural processes enters the technologist's laboratory and is called 'sustainability'.

If the global development scene is viewed as a test of sustainability rhetoric, we might well conclude that sustaining ourselves is paramount. This does not seem unreasonable, but it is a position that readily leads into sustaining economic progress and existing social hierarchies. Thus we ask *is our country sustainable?*[330] We then frame the question technologically within an assumption of continual economic growth. In agriculture, this logic justifies molecular manipulation of genetic material to suit existing or changing natural environments with the objective of ensuring continued profitability of farming. Why? Because *sustainability is not possible without profit*.[331] It is true that real profit is not possible without sustainability, but short-term profits certainly are. The implication of reliance on technology is also confused. Of course many technologies are beneficial. But to use sustainability rhetoric to justify investment in a technology is no more honest that claiming that the technology is antithetical to sustainability. As we will discuss further in a later chapter, technologies remain morally neutral. It is the motivation of their research and application that determines the morality of their consequences.

So, the barriers surrounding reductionist science and its education support systems[332] redefine sustainability to suit the consumer-based system of the West. This is hardly the *anthropo-transcedant*[333] attitude of species equity discussed by some broad-based scientists and evident in the secular and

---

[329] Seccombe (2000)
[330] Harris (2000)
[331] Peacock (2000)
[332] Watts (2000)
[333] Greenfield (2000)

theological arguments considered in earlier chapters. We may summarize the discussion as follows:

- Assumptions that global resource limitations would cause catastrophes have been revised by inclusion of our innovative potential, which has become a tenet of sustainability.
- Reliance on development of agricultural technologies to maintain economic growth easily neglects consideration of alternative non growth-based sustainability options.
- The continuous growth model that has been extended to less-developed countries has placed sustainability as a servant of economic growth and existing global hierarchies.

Our development rhetoric offers little to understanding the elevated aspects of sustainability discussed in earlier chapters, for it belies an unexpressed motivation to sustain some lifestyles ahead of others. Perhaps it is indeed true that hypocrisy is the mark of civilization. In any case, the bandying-about of sustainability in less-developed countries has produced some additional considerations of agricultural sustainability. These are examined further in the following chapter.

Chapter 15

Words versus Actions in Global Agriculture: Sustainability in Less-Developed Countries

When I am working on a problem, I never think about beauty. I think only how to solve the problem. But when I have finished, if the solution is not beautiful, I know it is wrong.

Buckminster Fuller

The word sustainability has been so variously specified, though never beautifully, that it is pointless to here postulate one more definition. Rather in this chapter we will briefly examine the word then look at our actions in less-developed countries. As the previous chapter suggests, actions may offer a window to motivations behind the word 'sustainability', which is increasingly considered as an economic good.

The common Western economic model describes its products as 'goods' but has omitted the reciprocal of 'bads' that were described in Adam Smith's original conception.[334] Is this why we so commonly focus on sustainability as a product? Economic analyses do acknowledge 'bads' as 'contingencies' but bury reference to them in text separated from recommendations on which decisions are based. It seems that 'bads' are viewed as tolerable by-products of 'goods'. We have in fact tolerated negative environmental impacts until recently, but now seem to want the 'goods' without the 'bads'. And we call this 'sustainability'. But while we pursue an

[334] Nicklin (2000)

objective at all costs without considering side-effects, we widen the gulf between the actions and rhetoric of sustainability. If the stock market offers no premium for ethical investments or triple-bottom-line accounting, it indicates our actions continue to accept the 'bads' as long as we benefit from the 'goods'. Expressed in other modern terms, we retain the 'right' to maximize personal financial profits while expecting others to curtail their profits to suit our aspirations.

We also invoke sustainability as a counter to catastrophes predicted from erroneous calculations, such as limited resources being exceeded by population growth.[335] It is worth noting that crises in less-developed countries usually stimulate major economic expansion through entrepreneurship, management and technology. It would appear that something is being sustained – is it our ingenuity?

Technology increases economic efficiency cumulatively and is easily transferred. Its economic contribution is equivalent, in economic terms, to an expansion of a resource base. Therefore, inclusion of technology in economic models can negate concerns about 'economic sustainability'. But within the economic paradigm, preservation of the environmental 'capital' account remains an imperative to interpret sustainable development as conceived in the Brundtland Report. Perhaps we should restrict our discussion to the indefinable 'ecological sustainability'.[336]

'Ecological sustainability' speaks of values and quality of life more than survival.[337] Each generation bequeaths a higher total capital asset to the next, including the natural environment. In this way sustainability is used *as a characteristic of systems; as a decision objective; for continued*

---

[335] Malthus (1914)
[336] Nicklin (2000)
[337] Nicklin (2000) Page 8

human survival; as dynamic human economic ecosystems, or *as a socially useful natural phenomena with socio-technical complexes.*[338]

Definitions of sustainability include 'sustainable human development' as a societal objective to meet the basic human needs and aspirations of future generations.[339] They also include sustainability as *a production system that indefinitely meets demands for food and fiber at socially acceptable economic and environmental costs [which is agreed to be ambiguous as environmental costs, unlike economic costs do not arise out of market transactions and are thus not priced]*. The list of definitions of 'sustainability' and 'sustainable development' extends to more than one thousand – and that was only to 1993.[340] Nevertheless, even though no unified operational principles have emerged, we can see three broad groups of definitions; economic, environmental, and socio-cultural.[341]

What has this meant in less-developed countries? Brundtland's *new concept of economic growth* indeed led to the replacing of definitions of sustainability to be *a new name for environmental protection.*[342] This produced the attitudes of having the cake while eating it as recommended in such papers as *Agricultural Growth, Poverty Alleviation, and Environmental Sustainability: Having It All.*[343]

Seeking growth and sustainability together is fraught with compromise. Global rises in population and incomes are projected to require continued increases in agricultural productivity, which all rely on technology. And it is not just the technology of recent years, for while potential yields continue to increase, the rate of that increase has now declined at the same time that availability of agricultural land is

---

[338] Shivakoti et al (1997)
[339] Malone (1994)
[340] Anderson (1993)
[341] Munasinghe (1993)
[342] Brundtland (1987) Pages viii-x
[343] Hazell (2000)

declining. So the problem for less-developed countries is defined as being how to maintain agricultural growth and alleviate poverty, while improving environmental quality. Hence the ideal agriculture is *broadly based and involves small- and medium-sized farms, and is market driven, participatory, decentralized, and driven by technological change that enhances factor productivity but does not degrade the resource base.*[344] We have not even come close to achieving such an internally-conflicting wish that ignores urban and population issues.

Rapid population growth in marginal agricultural areas with limited opportunities for off-farm employment creates future urban slum dwellers. Those who do not migrate to cities are forced to use more and more environmentally sensitive land for food production. Of course *environmental degradation [is] not an inevitable outcome of agricultural growth.*[345] It can be called misplacement of economic incentives. But we should ask where in the world have we got the incentives right. Using agriculture as a basis for sustainability may well be flawed from the start when demand for food seems elastic at times and inelastic at others. And we find that definitions of environmental degradation from agriculture refer to quite different things. For example, it is inevitable that a poverty-stricken farmer on marginal land will over-use his poor land to feed his family as his time perspective becomes shorter and shorter. Similarly, in a more-developed country, agricultural subsidies for intrusive land re-forming, incentives to over-produce, and excessive chemical use are now recognized as being ecologically damaging, yet are tolerated.

With this experience, definitions of sustainability in less-developed countries include – *a qualitative improvement without quantitative expansion beyond the widest definition of an ecosystem's capacity to regenerate the materials extracted by the*

---

[344] Hazell (2000)

[345] Hazell (2000)

development activity and the capacity to absorb waste products including energy of the development. Its achievement relies on a series of activities and institutions.[346] Each such definition is at least subtly influenced by the Brundtland Report.[347] In Australia, this was clearly the basis of *ecologically sustainable development,* about which many technologists were concerned that it gave unwarranted licence to precautionary decision-making that could block economic development.[348]

Such an economic development idyll is matched by the romantic poet trapped in the modern reductionist scientist who dreams of sustainable agriculture as a state where *salinity returns to its ecological niche, where deep-rooted crops and pastures tap the nutrients and water deep in the soil, ... where trees are encouraged to take their vital place in the environment, providing shelter, shade, produce, natural habitat and most importantly, a restoration of the balance between the underground and surface water supply [and] where cropping systems maintain the soil against wind and water erosion, particularly on the gentle slopes that are so badly abused.*[349] The stated aim in this case was *to find ways of maintaining and increasing plant and animal productivity while maintaining soil structure and fertility and without producing toxic residues in soil, water, fibre and food.* In these terms sustainability is an ideal - a distant target - rather than something accomplishable within the funding tranches of local governments.

But the commercial orientation of these definitions does not gel with the needs of subsistence communities in less-developed countries for it assumes that they must join the international monetary system. Growing cash crops in place of family food crops introduces risks greater than simple economic analyses indicate for peasants who integrate

---

[346] Lebel and Steffen (1999)
[347] Fitzpatrick (2000)
[348] O'Brien (2000)
[349] Campbell (1989)

agriculture into their overall lifestyles. An example was reported to the Brundtland Commission itself of the outcome of such policies; in 1984 five Sahelian nations experienced shortages of food in previously food secure areas because farmers had been enticed to plant cotton instead of food crops at a time when international cotton prices were falling.[350]

Policies in less-developed countries are inevitably influenced by the World Bank, which states that economic and environmental objectives are inseparable in the form of *maximization of net benefits of economic and social development subject to maintaining the services from, and stock of, natural resources over time.* It might work if renewable resources were used at a slower rate than their natural rate of regeneration and waste was created at a rate below the assimilative capacity of the environment. But in fact resource utilization and pollution are freely traded, which readily makes unsavoury actions exportable to poor countries.[351] Country policies are easily overrun by the potential short-term profits touted by the ever-welcomed foreign investors. All such pressures are part of international agriculture.

We may as well acknowledge that ignorance and population density will create further environmental impost, thereby rendering much of agriculture unsustainable in its present forms. Our actions suggest we have accepted this, for we tacitly tolerate environmental decline while population continues to rise. After the projected population has peaked, we can all be more sensible and sensitive.[352] However, this approach forestalls long-term preservation of an ecologically ideal situation.[353] Thus the issue of sustainability evolves.[354] We have reached the ridiculous situation of being seen to

---

[350] Giri (1984) in WCED (1987) Page 112
[351] Anderson (1993)
[352] Crosson and Anderson (1993)
[353] Lynam and Herdt (1989)
[354] Anderson (1993)

'sustain "sustainability" in development rhetoric' when, with a little wisdom, we could choose to see that insightful research has always focussed on sustainability without explicitly using the term.[355]

Wisdom implies a spiritual dimension that has been omitted from much of these discussions. Yet we now acknowledge that demand for environmental services, for example, rises once the basic needs of food, shelter and clothing have been met.[356] The observation sounds remarkably like those made by wise social observers 2,000 years ago. But in truth our current 'wisdom' is nothing more than a simple revision of the *agriculture as the engine of development model*[357] with economic growth encompassing sustainability values. This is hardly a giant leap from Brundtland.

The Brundtland Report overtly linked economic development to environmental protection in less-developed countries.[358] Today's interpretations of this may be paraphrased as follows - without alleviation of poverty, farmers must place the need to feed their families ahead of resource protection, and in any case, are ignorant of modern natural resource management knowledge. Thus the old bogeys reappear and impending catastrophe is forecast, beginning with a food crisis related to global soil losses and population increases, followed by exhaustion of cheap fossil fuels and then environmental imposts that exceed ecological thresholds and culminating in the demise of natural systems.[359]

Whether environmental apocalypse is likely or not, and I think it is not, the problem is not subsistence farmers in less-developed countries but their forced conversion to the cash cropping economy. Cash cropping may feed families better

---

[355] Anderson (1993)
[356] Braden and Kolstad (1991)
[357] Falvey (1999)
[358] Dutch Government (1995) Page 113
[359] Orr (1992)

that subsistence farming on average, but it introduces risks that generations of experience has widely rejected. And subsistence farmers are not necessarily poor within their own society. Subsistence agriculture should be considered of inherent value and a closer-to-sustainable form of agriculture than broad-scale agriculture. We tend to reject it as primitive just as we have for shifting cultivation, which in fact also appears to have been sustainable at low population densities. Our error is in assuming that commercial agriculture is always superior.

Defining Dimensions	**Agricultural System**		
	Subsistence	**Commercial**	**Sustainable**
Social identity	Family	Self	Community
World of reality	Past	Present	Future
Interpersonal processes	Conflict	Competition	Cooperation
Relationship to nature	Vulnerable to	Control over	Harmony with
Interpersonal relations	Mutual trust	Individual rights	Community needs
Natural Resources	Finite, consume	Develop, consume	Finite, conserve and preserve
Motivational drive	Safety and security	Self-achievement	Community accomplishment
Technological development	Borrowed or serendipitous	Faith as solution	Controlled for collective good

One approach to defining the features of commercial and subsistence (albeit in semi-utopian terms) agricultural systems is presented in the table.[360] It shows sustainable agriculture as involving community, being future oriented and

[360] Hatfield and Karlen (1993)

in harmony with nature, and taking a collective-good approach to the use of technology. The contrast to mainstream Western agriculture is so great that consideration of the approach is generally contained as a subset under *alternative agriculture.*

So we might conclude that our solution to date is not beautiful. Development rhetoric reveals a hidden hierarchy for sustainability to first sustain the relativity between rich and poor countries. The discussion can be summarized in the following:

- The diverse and conflicting uses of the word 'sustainability' are polarized around economic definitions that subjugate environmental contingencies.
- Influenced by Western thought, sustainability in less-developed countries is increasingly used as reassigned policies to stabilize population and reduce poverty through economic growth.
- Subsistence agriculture is overlooked in actions to convert less-developed country farmers to cash croppers, to the detriment of both the community and the environment.

So it might be said that, in exporting sustainability to less-developed countries, the West's actions have belied intentions to first sustain its own comfort. It offers little to our understanding of sustainability. On the other hand, the experience and traditions of less-developed countries may have something to teach the West about sustainability, as is considered in the following chapter.

Chapter 16

Learning Sustainability from Less-Developed Countries: Lost Traditions?

Years ago when you looked at a farmer's hands, they were like miniature lexicons of the landscape. The hands were worn and roughened through contact with soil and stone. Often rib lines of clay insinuated themselves into the lines of the skin. It was a powerful image of living hands remembering that originally they were, and would again be clay. People dressed in the Sunday best going to Mass. Serving Mass, you would see perfectly dressed men come to the altar for Holy Communion. They would stand reverently and offer this pair of withered, earthened palms on which the white host would glisten: the Bread of Life on hands of clay. This is a vignette from a vanishing world. Generally it seems that when we lose each other, we gradually lose our depth and diversity of presence. The world of function, instrument and image is a limbo where no presence lives, where no face is identifiable, where everything flattens into the one panel of sameness.

O'Donohue

The above extract[361] might well recall the perspective of a farmer in a less-developed country, but in fact is a memory from Ireland. While the West insists on sustainability being part of the approach of aid to poorer countries, use of the word sustainability is far from the above holistic view, as introduced in the previous chapter. Some might simply see sustainability as a Western intension or even ignorant desire to

---

[361] O'Donohue (1998) Pages 87-88

maintain its influence. Others might see it as a concern that balances the excesses of the materialistic system. In Asia, the tensions between millennia-old approaches and those of globalisation offer an understanding of the potential of sustainable agriculture in both less- and more-developed country situations.

But of course, the flow of global influence is biased by power. Regarding India, Tagore observed in 1908 – *we have for over a century been dragged by the prosperous west behind its chariot, choked by the dust, deafened by the noise, humbled by our own helplessness, and overwhelmed by the speed. ... If we ever ventured to ask, 'progress towards what, and progress for whom,' it was considered to be peculiarly and ridiculously oriental to entertain such doubts of the absoluteness of progress.*[362] More recently his countryman, Swarminathan observed that, *a world in which 20 per cent of the population enjoys 84 per cent of the annual income ... can never provide a secure and sustainable way of life for humankind.*[363] And before providing income, clothing, shelter, and medical care, food of course must be adequate. The Green Revolution was how agricultural science addressed the last of these four basic needs.

Setting aside environmental impacts for the moment, it is common for us to view the Green Revolution as a technological success. It provided food in less-developed countries and from this, maintained peace by averting war over food.[364] This was a real revolution as it supplanted existing forms of agriculture in less-developed countries. As a consequence it changed attitudes to nature, and to traditional agriculture which came to be seen as backward. It is revealing to compare modern myths about intensive mono-cultural

---

[362] Tagore (1941)
[363] Swarminathan (1994)
[364] Curtin (1995)

agriculture with alternatives from traditional or backward systems in the following table.[365]

The idea of 'permaculture'[366] is a current meeting point of these concepts in the West. It is similar to the traditional Asian smallholder, community-based, mixed farming paradigm and should be acknowledged as something much more insightful than 'organic' agriculture. As a new Western farming philosophy, permaculture seeks the mirage of sustainability and when applied in a considered manner can engender an awareness of cause-and-effect. With its high labour demands, diversity, small-scale, subjection to nature and so on, it would be described as backward in the above table. This makes it an accessible place for Western scientists to start to think beyond their training.

'Backward'	**Modern**
Labour intensive	Capital intensive
Diversity	Monoculture
Small-scale	Large-scale
Subjection to nature	Subjection of nature
Folk knowledge	Scientific knowledge
Generalists	Specialists
Integrated study	Reductionist study

'Backward' agriculture, in fact, seems to contain more of the attributes of sustainability than modern agriculture. Of course, the intention of subsistence farmers is usually to minimize agricultural risks rather than to protect the environment. Nevertheless, this prompts consideration of such seemingly outlandish alternatives as re-uniting with nature, as eschewing its commoditization. Again this implies that human's separation from nature is a source of un-

[365] Levins (1986)
[366] Holmgren (2002)

sustainability, expressed in such forms as over-consumption. This theme produced one NGO's conclusion that in the final analysis, *informed and enlightened citizen action is [the] only hope.*[367] But there are other 'hopes' closer at hand - we might better ask whether ancient technologies, even shifting cultivation, offer a practical understanding of sustainable agriculture in the face of today's food needs.

Shifting cultivation systems ultimately founder as population density rises - yet they show a respect for natural cycles that has informed the integrated agricultural systems of Asia. The ensuing integrated cropping, animal husbandry and forestry agro-ecological approach continues to meet the needs of local peoples more readily than commercial approaches. It demonstrates techniques such as: crop interplanting in accordance with micro-ecological suitability; incorporation of agroforestry according to tree architecture to balance the canopy with the overall requirements of the mixed cropping and forestry system; encouragement of fast growing species to retain soil and reduce wind; use of valley areas for home gardens, and so on.

The integration of animal husbandry in farming systems facilitates nutrient recycling, and can be combined with technologies to produce biogas. Two other outcomes of such integrated systems are a sacredness afforded to aspects of nature and the encouragement of artisan skills based on local resources.[368] While both seem to be universal in traditional agricultural discussions in populous Asia, it is telling that neither is common in Western conceptions of Asian agriculture.

Sacred groves that have preserved remnant forests stand as monuments near landscapes decimated by non-traditional agriculture. They testify to alternative approaches

---

[367] ANCARD (1993)
[368] Ramakrishnan (1992)

to balance in the agricultural and natural ecology. Such groves, in Meghalaya in north-eastern India for example, are often located on hillsides and retain a biological diversity that is useful in regenerating other degraded areas. Describing these as *a good example of the sociological basis for nature conservation,* Ramakrishnan argues that such traditional practices also offer educational resources. Respect for traditional beliefs about the groves, uncommon in past development strategies, offers scope for sustainable agriculture in an interesting mix of new technologies and ancient practices, and even beliefs.[369]

Of course *religious concepts such as the soul, spirituality, or wisdom have been seen as unsuitable for scientific research.*[370] Even today they are but *fringe topics* in psychological research.[371] However, wisdom is an essential component of sustainability, as discussed in earlier chapters. We act wisely when we *attempt to be in harmony with an unseen order of things,* as James defined religion.[372] Traditional societies that find a harmony with their environment see untraditional acts as unethical and blasphemous. We may recall that this mimics the conclusion of an earlier chapter where 'sin' was defined as acting against the natural order. In such traditional societies, the leader is one who is in touch with the unseen order and who leads, acts and instructs all within that order.[373]

If we seek to change agriculture while blind to natural flows, we risk working against our espoused objectives in an endeavour such as sustainability. Having lost traditions, the West has become blind to the non-rational actions of traditional agriculture even when they point to critical components of sustainability. We may summarize the discussion as follows:

---

[369] Ramakrishnan (1992) Pages 390-394
[370] Levitt (1999)
[371] Chandler and Holliday (1990)
[372] James (1994)
[373] Peck (2001) Pages 46-48, 267

- Attitudes to nature in the traditional agriculture of less-developed countries contrast with those of modern agricultural approaches and support the contention that sustainability in agriculture is inseparable from overall attitudes to nature.
- Religious conventions and traditions protect nature against human excesses in less-developed countries, even though such countries may appear to Western eyes to have no productive advantage within a specific agricultural enterprise.
- Small-scale agriculture as practised in Asia provides a basis for expanding the concept of ecological sustainability to include agricultural ecosystems.

Traditions lost from everyday Western approaches yet retained in less-developed countries show us a link between the spiritual, religious and technological aspects of agricultural sustainability. This understanding, afforded to a few in each recent generation, has encouraged the emergence of a spirit of sustainable agriculture among some practitioners in more-developed countries, a theme which is amplified in the following chapter.

Chapter 17

The Emerging Spirit of Sustainable Agriculture: Changing the Western Worldview?

Those who fall in love with practice without science are like a sailor who enters a ship without a helm or a compass, and who never can be certain where he is going ...

da Vinci

Rather than extend its version of sustainability to less-developing countries, the West would do well to learn from those countries. Small-holder agriculture more easily balances spiritual with production aspects than large-scale intensive agriculture where commercial and environmental objectives often conflict. When one stops to notice that modern agriculture seeks to control as many production variables as possible, one can immediately see a compromise with sustainability. The choice is whether understanding and interacting with the natural environment is sustainability, as I assume in this manuscript, or whether sustainability is an ever-rising ability to control natural processes. As with all views, such extremes are only useful as discussion points, and then only when considered without subjective attachment. In this case, the discussion highlights the benefit of understanding more of peasant value systems in such regions as Asia.

A recent Asia-wide review defined sustainable agriculture as the sustained ability to provide adequate food supplies[374] - a definition popular in countries that benefited

[374] Kaosa-ard and Rerkasem (2000)

from the Green Revolution. That revolution was followed by interventions of the Asian Development Bank, which defines sustainable development as *that which can evolve indefinitely toward greater productivity and human utility, enhance protection and conservation of the natural resource base, and ensure a favourable balance with the environment*. The economic orientation of the terminology orients sustainable practices to a balance of natural resource replenishment with rates of consumption. No mention is made of the role of traditional religious values or morals in this balance - it remains overwhelmingly an economic goal.

Re-investment in the resource base is interpreted to mean technological research that effectively increases the utility of resources and *facilitates the exploitation of nature*.[375] And technological management is indeed critical for modern agriculture to solve problems such as; pest control, nutritional deficiency, irrigation-induced waterlogging and salinity, and reduced genetic diversity. In addition to technical issues are those that are grouped together as 'governance failures'. Governance failure occurs in such forms as; environmentally damaging projects, misguided sectoral policies, inequitable land and water policy and institutional weakness. So what would happen to the major food stuff of the world rice, for example, if this approach were followed as stated? Well, the very analysis referred to suggests the demise of rice unless technologies and policies that enhance its 'sustainability' are implemented.[376] It behoves us at this point to recall that rice-agriculture appears to have been one of the sustainable forms of agriculture before use of the intensive production model.[377] The narrow logic that sees a simple transplantation from Western systems to Asia easily ignores such past Asian experience. And acting in ignorance always produces

---

[375] Kaosa-ard and Rerkasem (2000) Page 136
[376] Kaosa-ard and Rerkasem (2000) Page 198
[377] Falvey(2000)

undesirable outcomes. If we are to accommodate the element that we feel is missing, we must turn to other more sensitive analyses, such as that of the British colonialist who left us *An Agricultural Testament*.[378]

Sir Robert Howard, once director of an Indian research institute, describes his epiphany in appreciating the natural integrative systems of Indian farmers. In the 1930s, he became concerned about the impact of industrial approaches to food production in terms of lost soil fertility and an increasing natural imbalance indicated in disease and erosion. Considering soil as the earth's capital, he contrasted the durability of the natural Indian humus-based agriculture with the demise of past civilizations that lost their respect for soil. He also argued that *at least half the illnesses of mankind will disappear once our food supplies are raised from fertile soil and consumed in a fresh condition*.[379] We now know that his scientific explanations of the actions of humus were flawed, but his main principle still feels correct – the health of soil is related to the health of people working with it and consuming its food.

So how can we draw these experiences together? Development assistance in Asia provides a clue when it acknowledges systems approaches – that is, approaches that acknowledge interactions with the many factors affecting outputs. Systems approaches, however, are often avoided by the reductionist scientists and the influence of such refinements as Participatory Action Research[380] is vastly less than that of industrial agriculture. Participatory research attempts to incorporate community values and external ideas into the creation of a knowledge base for project planning. Of course, participation in this case does not extend beyond humans, so that biological integrity, for example, is subservient to the value systems of government and funding

---

378 Howard (1943)
379 Howard (1943)
380 Leake (2000)

agencies. Once again, we seem to have come close to the essence of sustainability only to veer away.

The link between soil and human health has, of course, been made by many others, including Balfour, whose thesis was tested experimentally over thirty years. She suggested that unintended changes of single-factor research impacted soil health. Changes in micro- and macro-organism balances were accompanied by increased need for pesticides, which she saw as affecting human health.[381] In the pragmatic 1940s, this approach omitted any mention of spiritual interactions with the natural environment - these had to await our own age. We now licence ourselves to describe, for example, agriculture as a self-sustaining garden within the natural environment, providing produce and reproducing itself through time, and in which *the true gardener gardens himself* as he practices *the most demanding of all the arts*.[382] This is a Western reclaiming of deeper values in its interaction with the natural world. Who knows - perhaps the family farmer will again one day be recognized as a spiritual person.

Spiritual approaches to agriculture are far from being the antithesis of intellectual insights. The separation at fault is that between intellectual understanding and technical applications. We may consider the natural link in the case of Goethe's intellectual insights being translated into practice through Steiner's eight lectures on agriculture.[383] These intriguing and sometimes frustrating lectures unite divine elements and agriculture, linkages that have offended scientific thought yet have sometimes foreshadowed scientific discovery. For example, Steiner postulated the absorption of complex organic molecules by plants when scientific convention held, until recently, that only inorganic elements were absorbed. Other premises of Steiner are worthy of

---

[381] Balfour (1975)
[382] Hodges (1986) Pages 114-115
[383] Steiner (1974)

recognition, including natural fertilizer techniques, the balance between plant health and disease resistance, the linkages between the cosmic quality of food and human health, and the impoverishment of food as a function of industrial approaches to farming. Each of these can appear as misguided to purists as 'organic' produce. Yet their overall context reveals a commonality with the integrity of life and life-supporting forms that would be familiar to an Asian peasant.

Steiner's anthroposophy built on the earlier 'biodynamic agriculture', which remains separated from modern and even organic agriculture by its *awareness or sense that every living being has a link with the spiritual cosmic world, and that it is the duty of every human being to guide the life of these beings in such a way that these links can take place undisturbed.*[384] By treating a farm as a living organism, value is given to the therapeutic and wide social benefits of working within natural cycles rather than according to market prices. But any modern environmental prophet saying this is a voice crying in the wilderness. He may even be misinterpreted as seeking to re-create wilderness from farmed land. But the West does have both the information and the spirit for a sustainable approach to agriculture within its community. It requires that all factors, from the technical to the psychological, are acknowledged in an integrative style similar to that for which the profession of agricultural science was once famous.

We may summarize the discussion in this chapter as follows:

- The uniting of spiritual and intellectual aspects of agriculture as in less-developed countries is usually ignored in favour of Western approaches that treat sustainability in economic terms devoid of even considered intellectual analysis.

---

[384] Schilthuis (1994) Page 55

- Adding technology to the production environment as a further resource in a manner similar to the commoditization of natural resources may possibly stimulate reconsideration of integrative approaches and humanness.
- Linking of individual and societal health to holistic agricultural practices continue to haunt science.

But we should not despair, because Western culture still retains some links between agriculture and spirituality, connections that remain consistent with peasant agriculture. Current thought of these links warrants our further consideration, as is done in the following chapter.

Chapter 18

Unity in Diversity of Views? Spirituality in Modern Agriculture

It is the way of Heaven to take from those who have too much and to give to those who have too little. But the way of man is not so. He takes away from those who have too little, to add to his own superabundance. What man is there that can take of his own superabundance and give it to mankind? Only he who possesses Tao.

the Tao ti Ching

The improbability of the West fully understanding the indigenous knowledge systems of other cultures has confused much of its understanding about alternative agriculture. It has led to modern agriculture trying indigenous practices from less-developed countries isolated from their values, with the effect that unforeseen outcomes cause such practices to be discarded as impractical. From the opposing camp, romantic views of traditional agricultural systems usually ignore their dynamism and constant accumulation of knowledge, even sometimes within government. For example, modern China insisted on allocating land to small-holders despite Western advice to consolidate into large holdings suited to industrial agriculture. The result has been that China has fed itself and has adapted modern science to traditional approaches. But then, we must ask, is this sustainable?

There is little doubt that Chinese agriculture, at least until recently, recycled natural wastes more efficiently than Western approaches, but this may have been a function of population and necessity. Of greater significance is the current very high chemical fertilizer use of Chinese agriculture, [385] which now contaminates ground and surface water.[386] China uses whatever technology can be applied within its socio-economic system, and it would be naïve to suggest that technology is the sole descriptor of such agriculture. It may not seem sustainable, but it is meeting the overall need to feed a billion people and to minimize urban immigration. Thus the system has proved to be resilient – and resilience may be a component of sustainability. But China offers more limited scope for understanding the role of religion in agriculture than other Asian countries, the most populous of which is India.

Indian traditions surrounding the natural world stem from *precriti*, the Sanskrit word for nature which includes such factors as the inner drive of all life, creativity, fertility of the soil, and renewable water sources.[387] This respect for the self-organization of nature provides a perspective on the divine that contrasts with concepts of a single creator God who is increasingly marginalized by technological innovations such as genetic modification. It acknowledges that traditional agro-sociological systems are an integral part of religio-cultural observance. Thus the sacredness of the Indian cow allowed for milk production, traction and draft, fuel, and fertilizer, and in some cases, leather and other products when the cow died – a stark contrast to the single-output approach to bovines common to Western systems.

The Indian system is often defined as of low efficiency when in fact it is the high costs of artificial feeding regimes and pollution from intensive factory farms to produce a single product that have low net efficiencies in terms of resource use. The role and value of multi-purpose draught bovines compared to tractors in less-developed countries has long

---

[385] Huizer (1995) Pages 31-33
[386] Zhu and Chen (2002)
[387] Shiva (1995) Pages 38-42

illustrated the point.[388] Promotion of Western technologies in less-developed countries can easily ignore local efficiencies that may well point us toward sustainability. This does not mean, as some romantics claim, that peasant agriculture is all good and intensive agriculture is all bad. But such sentiment is readily fuelled by actions of Western groups who insist that hi-tech systems are sustainable and should be introduced to less-developed countries while restricting access to technology through such acts as patenting of genetic material.[389] If motivation is the key to an action being sustainable, then profit motivations conflict with espousals of sustainability.

Genetic modification, which allows such novelties as human milk and pharmaceutical proteins secreted from domestic animals, has moved agriculture one step closer to becoming a factory. While it misses the point to unquestioningly label such developments as unethical, it serves to highlight the rights of animals to a life of minimal suffering. It also misses the point to argue that patents should refer only to new creations, and that as a consequence, modified genetic material should not be patentable. In theistic terms, it is seen that such *patents are a statement of man as creator*.[390] In the non-theistic terms of Buddhism, for example, the motivations behind such patents would determine their ethicality. Of course it is clear that no one scientist or group develops a new biological product. Even in our most insightful moments, we are as Einstein observed, always standing on the shoulders of the giants who came before us. Nevertheless, in debates about biological patents it is easy to confuse ethics with politics, rhetoric with truth, and fancy with reality – and such debates ultimately involve the Western-influenced elite of less-developed countries talking with or against the West. These debates are far from the spiritual aspects of agriculture,

---

388 Falvey (1988)
389 Shiva (1995) Pages 38-42
390 Shiva (1995) Page 42

yet as introduced in the previous chapter, they continue within Western culture and will sooner or later receive serious consideration.

Popular thought suggests that traditions of caring in Western agriculture have been progressively eroded since the European Enlightenment. The view is supported by the inclusion in the 1624 publication of Bacon's book *Nova Atlantis* of predictions of transgenic outcomes from human perfection of nature. In contrasting such a pronouncement with the 1500 trial of miners accused of killing mother earth, it is argued that the Baconian view has pervaded agriculture into modern times, leading to *synthetic agriculture: intensive, highly specialized, industrialized.*[391] It does not follow that such developments as soil-less crop production, controlled atmospheres, and measured provision of water must exclude spiritual relations with nature. Sure, the dynamism of nature is constrained, but the constrainer is part of nature too - there may well be a spiritual input and benefit to, for example, a hydroponic garden. The self-regulating systems of nature continue to operate in all cases and it is those who are agro-ecologically aware that will understand more of such systems. A spiritual association with agriculture in a true partnership with nature yields a self-realization and respect for all forms of life. This is demonstrably easier in systems that minimize their impact on nature, yet is not excluded from other approaches when they are essential and conducted with wisdom.

The spiritual element of agriculture acknowledges human needs to share life with plants and animals. This renders agriculture *fundamentally different from production of non-living products such as cars or radios.*[392] An understanding of ecological processes allows the culturing of a product only when it is a moral action, that is, it is cultured within natural

---

[391] Verhoog (1995) Page 44
[392] Verkleij (1995) Page 59

flows. Such an approach is seldom considered in industrialized agriculture, which can easily exceed ecological capacity to 'sustain' a system. Seen in this way, industrial agriculture might be just one indicator of the neglect of the *vestigia dei* in all beings and the *imago dei* in humans in the conception of Aquinas. Perhaps this is so. Some also argue that medieval conceptions of a united cosmos might mollify the excesses of modern technology itself. The point here is the attitude of practicing agriculture rather than the technology. To practice agriculture cooperatively within nature is to respect the divinity in all things.[393]

If it is true that each culture has difficulty acknowledging other cultures, the West's difficulty in understanding less-developed countries may be compounded by its weak spiritual bond with nature.[394] Yet obvious exceptions occur at the level of the individual. And is seems we can group individual approaches to nature into six attitudinal types;

- the despot who dominates nature without moral scruple,
- the enlightened ruler who reigns over nature while respecting it,
- the steward who manages nature on behalf of, for example, God,
- the partner who works equally with nature,
- the participant who conceives humans as one part of nature, and
- the *unio mystica* who has lost selfness and becomes one with the immanence of nature.

The despotic approach includes the West's modern technological orientation and its predecessors since the sixteenth century, notwithstanding occasional tempering through enlightened leadership and stewardship. This history

---

[393] Verkleij (1995) Page 62

[394] Zweers (1995) Pages 64-83

has produced today's definition of sustainability as the continued ability to fulfil human needs on behalf of future generations, or in some cases, God. To move to the stage of partnership or participatory relationship with nature requires revaluation of any assumed place above it. This is a difficult transition when the *modern, mathematical-mechanistic-technocractic worldview* determines actions, even to the extent of limiting effective development, as illustrated in the following example about plant biotechnology.[395]

A report on the ethical aspects of biotechnology in plants in Holland[396] concluded that technological advances may easily outpace moral debates such as simplistic representations of increasing food production to feed the rising global population. The report focused more on ethical, technological, and historical than spiritual perspectives to guide national policy, yet it offered a sensitive and sensible perspective uncommon in public policy. For example, genetic manipulation of plants may best be understood as a continuum of human developments over the last 10,000 years or so.

If we view plant breeding as beginning with the inadvertent influencing of plant characteristics by prehistoric humans, then modern molecular biological techniques are simply an acceleration of that process. Yet such biotechnological advances are labelled as having negative outcomes even when they reduce reliance on pesticides and susceptibility to frosts and drought, or increase yields and suitability for processing. They also allow the aesthetically pleasing appearance of fruits most consumers demand as well as external-chemical-free production of human medicines. Risks no doubt exist. These are usually elicited in such forms as: ecological disturbance from genetic escape into wild

---

[395] Zweers (1995) Page 82

[396] Dutch Government (1995) Pages 84-116

populations; narrowing of genetic diversity; expansion of areas suited to agriculture; multinational ownership of genetic material, and poor country subservience to Western technology. But these are risks of conventional plant breeding also. The arguments have been elucidated in more detail elsewhere.[397] Ethical debates are poorly informed, even in Europe where one might expect public concerns to encourage education on the issues. We should also be aware that commercial multinational protagonists generate much of the debate's scientific information.[398] But our task is to relate these matters to sustainability and religion.

When religion offers absolute rules it fails such debates. One side may choose the higher moral ground of feeding the needy while another may value protecting the environment for all future inhabitants. Behind such stances may have been some insightful view that is lost when it is unthinkingly repeated, perhaps even making arguments illogical. For example, arguments about the intrinsic value of nature demanding minimal interference with natural cycles may be used against past plant breeding as well as against modern molecular techniques. Yet the answer is not that expressed in one multinational biochemical company's (ICI) study of allocating consideration of intrinsic values to religion and those of genetically modified crops to politics informed by science.[399] At the very least, religion needs to be informed by science and science by religion. But more than this, separation of any matter dealing with wholeness opens the field to vested influence.

Self-interest is of course not the preserve of commerce. Religion in its cultural forms acts in the same way. As we have discovered earlier, institutionalized religion is poorly equipped to lead any debate on environmental matters for its

[397] Falvey (2004)
[398] Straughan (1992)
[399] Straughan (1992)

spokespersons are often poorly informed. But science and politics are also poorly equipped for moral matters when we think in terms of the wider human capacity for wisdom above knowledge. The Dutch report referred to above offers a practical pluralistic ethic when it observes that *few values are so sacrosanct that they never have to make way for other values [and] our concept of nature is about to change.*[400] The shift it refers to is one from the anthropocentric worldview prevalent in the West since the Renaissance to a more eco-centric worldview.

A shift in worldview to respect ecological functionality might be thought to favour local production and consumption of food and small-holder agriculture. But it does not necessarily mean this nor does it mean no change in existing ecosystems. When we recall that plant breeding began with the accidental germination of edible seeds around ancient settlements, we can see a continuous human modification of ecosystems. Plant breeding developed further as superior seed was chosen and transferred to new environments and when cross-breeding was found to enhance desirable traits and produce hybrid vigour. The totipotent (able to generate a complete organism from a cell) character of plants allows reproduction to be artificially induced from any plant part and has allowed advances ahead of animal science.

Modern plant breeding has effectively separated plants from many environmental constraints. We can theoretically breed plants suited to most inhabited environments. So will this mean that each part of the world can produce its own food? Theoretically yes; but then most parts of the world do anyway. And most of the world's farmers are small-holders. It remains moot as to the extent that the changing Western environmental ethic will be able to effect much change from the current situation unless something more than environmental concern informs this zeal. Surpassing technical

---

[400] Dutch Government (1995) Page 91

matters, it is inequities in global trading and sharing of food surpluses or deficits that produce continuing problems. Far from being a failure of nature or agriculture to provide adequate food, this represents individualistic attitudes at work - attitudes that allow one to ever indulge oneself before thinking of sharing with those whom one does not know.[401]

	Dominator	**Steward**	**Partner**	**Participant**
Herbicide resistance	Yes	Yes	Perhaps	No
Disease resistance	Yes	Yes	Yes	Perhaps
Stress resistance[a]	Yes	Yes	Yes	Perhaps
Aesthetic change[b]	Yes	Perhaps	No	No
Alien genes	Yes	Perhaps	No	No
Ownership by patents	Yes	Perhaps	No	No
PBR[c] protection	Yes	Yes	Yes	Perhaps

[a] salt, drought [b] taste, colour, shape [c] plant breeders rights

So our ethical behaviour belies our rhetoric of equity and environmental integrity. We discuss subjects in terms of relative ethical acceptability. If this approach classifies such attitudes to genetic engineering into acceptable, unacceptable, or conditional, different views may be tabulated as above[402] by using four of the six attitudes toward nature listed as bullet points earlier in this chapter.[403] The approach is limited yet useful in highlighting the question – what is a participant with

[401] Dutch Government (1995) Page 95
[402] Dutch Government (1995) Page 107
[403] Zweers (1995) Pages 64-83

nature? To address such a practical question in religious terms requires more than the Christian orientation of the Dutch studies and will be informed further in subsequent chapters that consider Eastern insights.

Such analytical approaches assist decision-making in an informed society like the Netherlands. It may be applicable elsewhere. From our viewpoint, it is sufficient to note that the description of participation with nature is similar to that of traditional subsistence systems of agriculture, which wherever they remained sustainable, were part of religious systems.

Before developing our religious theme further, one further comment on the Dutch work may be useful. It is clear that ethical evaluation of plant genetic modification cannot be limited to extrinsic consequences. In addition, we should be cautious in use of the argument of the 10,000 years continuous plant breeding when we note that mutagenesis and transgenesis are apparently rare in nature. The intrinsic value of nature is not considered in the self-limiting approaches of science. And Western culture separated from spiritual insight relies heavily on the rational interpretations of science. But in the underlying experience of the insightful practitioners we find a deeper form of knowledge. Rather than seek to unravel the rational from the experiential to reinterpret philosophers and sages revered in the West, we can examine some Eastern insights on nature that have not been totally rationalized.

So this discussion brings us back to the lessons of peasant farmers and spiritual involvement with agriculture. It may be summarized as follows:

- While we cannot expect to understand the full operations of a traditional small-holder agricultural system, we can glean a perspective of its integrity which indicates that agriculture cannot be defined solely by technology.
- Whatever religious metaphor is employed to convey the intrinsic inter-relatedness of nature, it assists whole-of-

system understanding that is critical to sustainability in a manner uncommon in technological discussions.

- Although the spiritual aspects of agriculture have been separated from its science, they seem to have survived in smaller scale agriculture and gardening in which participation in nature is balanced with output.

The next chapter introduces a wider perspective of religion as a precursor to consideration of Eastern experiences in the subsequent chapters.

Chapter 19

Bridging the Break: Reconnecting Through Religion

A man will reach perfection if he does his duty as an act of worship to the Lord, who is the source of the universe, prompting all action, everywhere present ... for you yourself have created the karma that binds you. You are helpless in its power but you will do that very thing which your ignorance seeks to avoid.[404]

Bhagavad-Gita

Notwithstanding the preceding discussion, science is part of the disaggregated message of modern Christianity and as such may offer a means of integrating spiritual and secular knowledge. Yet to see the integrated whole, we must rise above the mire of religious history that is the antithesis of the reunification implicit in the Latin *religio* – reconnection. In this case, we may see this as reconnection within nature.[405] But the very tools we use are tools of alienation, of separation. If the evolution of language facilitated community cooperation, it must also have contributed to the diversity of relationships that defines humanity in terms of art, thought, culture, and abstract thinking. And as our abstract thinking became more complex, it assumed a reality that has alienated us from nature.

---

[404] Prabhavananda and Isherwood Page 158

[405] Capra (1996) Pages 285-288

The tension between our yearning for a lost integration with nature and our fragmented lifestyles has long consumed artists, poets, philosophers and mystics. It seems to be the same focus as *suffering* or *unsatisfactoriness* in Buddhism, which Capra considers to *contain some of the most lucid expositions of the human condition and its roots in language and consciousness*.[406] This chapter therefore seeks to be a simple bridge between Western and Eastern approaches that can lead us across the essence of Buddhism in the language of sustainability unencumbered by dogma and tradition.

The once *creative spiritual vision [that] gradually ossifies during its transmission through the ages*[407] still pervades all great religions and forms a common basis for discussion. Yet this personal experience that is the intent of the pre-ossified teachings has all but proved impossible to convey in mass terms, and remains uninteresting to most persons. Let us revert to Western culture for an example. Socrates' rationalist approach may be interpreted to lack an understanding of the harmony of nature, whereas Jesus may be seen to have been in harmony with nature in his life and teachings. But Christianity grew out of Paul's more than Jesus' efforts, and Paul like Socrates missed the balance of all things, thereby instilling his legalistic approach in the subsequent Christian interpretations. Through Paul's interpretation of Christ's life and death we have inherited a narrow definition of salvation through faith rather than by the personal spiritual development supported by an ever growing confidence, called faith, in the model afforded by Jesus' life. Similarly, Paul's interpretation of the *son of man* as the *son of God* introduced the second person of the divinity when this may well not have been Jesus' intention. Such actions by Paul then led to re-interpretation of the

---

406 Capra (1996) Page 286
407 Subuti (1995) Page 45

Hebrew-Arabic ritual of sharing of food and drink into transubstantiation in Christianity.[408]

The above example is a succinct and apposite example of religious convention being an unreliable guide to the spiritual intent of these great prophets. We may relate this to agricultural sustainability in such simple ways as the West's separation from the cycles of agricultural feasts. This is why the Quakers – the most Buddhist-like of Christian groups – acknowledge cycles. They also reject the hierarchical priesthood and institutionalized doctrines of the church.[409] When we realize these things, we understand how the emphasis on stability, sustainability and security may be but a form of idolatry within modern Christianity. We may also see that messages lost from the West are now being re-learned from Eastern traditions. We can see this in terms of a return to; seeking God within ourselves or in all beings, considering God to be an experience rather than a defined being, and understanding prayer as meditation rather than a request to a more powerful figure.[410] From that Oriental exposure, we can see that Christian respectability surrounding rationally conceived sustainability has carved a new idol.

Modern scientific progress has developed without significant interaction with the East until relatively recently. And understanding of Eastern philosophy in the West remains poorly developed outside of the very few who have immersed themselves in the traditions. The oldest of sacred literature dates from 1500 to 1000 BCE in the form of the Vedas of India, which take the form of praise of the divine and protection from evil. It was the revolution leading up to 500 BCE (that incidentally coincided with the religious advances of Palestine and Greece) that produced the Upanishads of India in which the personal and universal forms of Atman in humans are

---

[408] Han (2000)
[409] Cupitt (2000)
[410] Spong (1998)

complemented with a unified Brahman as God, with the two being combined as one at times. The Upanishads contain the concepts of the non-existence of a separate self and of immortality, both concepts critical to an informed understanding of sustainability. They also question the existence of the duality upon which human rationality is based, and highlight the hindrance that duality presents to insightful understanding.[411] If we think on the fundamental duality that is linear time, then we might claim that the Upanishads first explained the irrelevance of such human conceptions as sustainability.

Possibly around the same era as the Upanishads, the Chinese Tao presented the undivided unity of all things as only fragmented by our reliance on duality in such opposites as yang and yin, and dark and light. To understand the Tao is to live a moral life that follows the nature of the cosmos. So, the Tao attracts its advocates as an explanation of the excesses of humans in modern agriculture for example, or more popularly in seeking to justify our sensitivities to natural pollution. But the Tao is more appropriately understood as a description of the nature of all things - a concept also expressed in the Indian Bhagavad-Gita composed between 100 BCE and 200 CE and which expanded the concept of God and human actions in terms of love and wisdom.

Such early documentation of insights points to our forebears' gradual realization of the nature of reality, both as a curiosity and as an explanation of the difficulties encountered in the ordinary life when we seek to work against reality. The evolution exists in the Judeo-Christian tradition, though it is often clouded by forced unification of Old and New Testament insights. The West has now embraced a new means of understanding this evolution through exposure to Oriental traditions, most commonly in the form of Buddhism - and it is

---

[411] Happold (1991) Page 148

Buddhism that is increasingly, and usually erroneously, used to advocate environmental responsibility.

The Buddha has been claimed as the first environmental thinker on the basis of isolated references from the voluminous Buddhist scriptures.[412] But the scriptures usually use environmental examples to illustrate spiritual teachings rather than as guidelines in themselves. For example, the teaching, *how astonishing it is that a man should be so evil as to break a branch of the tree, after eating his fill from it*[413] points to the interdependence of all things by using an everyday environmental story. To interpret it otherwise misses the central point of Buddhism as captured in the popular Zen saying, *when the master points at the moon, the fool looks at his finger.* Ancient references to forests possibly reflect a concern for the environment, but should be read in their spiritual and historical context, which would suggest their concern with harmony, compassionate lifestyles, or as a reaction to the rising urbanization and agricultural expansion of this time in India. Likewise, Jataka myths of the Buddha's lives as animals are less environmental teachings than they are cultural reassignments of traditional stories to illustrate wisdom, non-violence, compassion, and loving-kindness. Causal dependence, the embracing insight of Buddhism (refer to Chapter 23), does offer ready parallels to ecology, but it is much more - it is an overarching operation in our psychology and all other things.

Apart from projections and misconceptions, the West's encounter with the East through Buddhism offers us an understanding of motivations in seeking sustainability. For example, Buddhism explains that the world is governed by laws of cause and effect, and interdependence. It is not governed by a God who, according to the logic of Buddhism,

---

412 Palmer (2001) Pages1-7
413 Anguttara Nikaya Vol III:262

could hardly be considered omnipotent, eternal and autonomous after indulging in the act of creating an ongoing world.[414]

The insights of Buddhism applied to modern approaches may see Western science and technology as a *major contribution to minor needs*[415] - though Ricard may have subsequently softened his view on this.[416] Nevertheless, most of modern science addresses issues that are superficial compared to spiritual development. We may reconcile this with modern viewpoints when we realize that science was once the ground of philosophers before it shifted to its current technological orientation. The shift impoverished Western philosophy as it shed the ancient richness of philosophers who lived according to the morality they had proved personally through reflection, logic and debate.[417] It is this separation of science from philosophy that has produced today's amoral science, which in agriculture is proving unable to distinguish between beneficial progress and the mere application of new technologies in the factory farming of domestic animals, for example. One can see a clearer logic in Buddhist arguments of cause and effect in the links between new human and animal diseases, such as 'mad cow' disease. In fact the main clouding of such insights is the assumption of constant progress - a belief that we can always develop a solution.

Aldous Huxley uncovered the sandy foundation of Western belief in continuous progress when he noted that we value novelty in technologies and art above attentive reproduction of ancient techniques, adherence to traditional modes of technology, and even morals.[418] We seem to have abandoned the introspective detachment of ancient Greece

---

[414] Revel and Ricard (1998) Pages 116-118
[415] Revel and Ricard (1998) Pages 134-135
[416] Matthieu Ricard (pers.com.)
[417] Revel and Ricard (1998) Pages 185-199
[418] Huxley (1947)

that provided the contextual linkage between knowledge and wisdom. Of course, it may be argued that many ancient beliefs have proven erroneous, but we might better see such erroneous bits as the bathwater of knowledge rather than as the baby of wisdom.[419] Even the word 'ritual', which is derived from the Latin *ritus* or 'correct action' speaks of the same message that is now entering the West in the form of 'right action' in the Dharma of the Buddhist Eight-fold Noble Path. To me it is no mere coincidence that early uses of the word *Dharma* may be interpreted as 'duty' in the sense of correct action in accord with nature in its broadest definition.

Both ritual and Dharma in these contexts point to the need for our continuous development of insight. In the Christian conception, the opposite of such correct action defines 'sin'. And because it *turns consciousness away from truth [as] the consequence of illusion and selfishness, sin [thus] includes its own punishment.*[420] In terms of sustainability, correct actions are those which accord with the natural order and are thus sustainable, while incorrect actions produce unsustainable outcomes that a proper use of Christian language might call sinful.

In linking East and West, this chapter may be summarized as follows:

- The reconnection implied by 'religion' may be assisted by the interaction of Eastern and Western spiritual traditions, using environmental considerations as their current popular dialogue.
- The progressive evolution of our self-knowledge as a spiritual action provides a perspective on ancient scriptures, one that can reduce their erroneous literal

[419] Revel and Ricard (1998) Pages 281-300
[420] Revel and Ricard (1998) Page 238

application to modern sensitivities in environmental discourse.

- Buddhism offers one means of examining sustainability in a manner that accepts scientific discourse through its basis in cause and effect from which moral guidelines are drawn as correct actions; the opposite is sin in Western traditions.

The role of Buddhism in explaining the West's traditions, including its infatuation with sustainability, invites us to further examine the essence of Buddhism, which is the subject of the following chapters.

Chapter 20

Communicating with the Orient: Eastern Sustainable Agriculture

The Buddha gave his doctrine to enlighten the world: Christ gave his life. It is for Christians to discern the doctrine. Perhaps in the end the most valuable part of the doctrine of the Buddha is its interpretation of his [Jesus] life.

Whitehead

Is it true that 'Eastern' teachings may not be fully comprehendible without a detailed historical, cosmological, sociological and psychological understanding of the various spiritual investigations that have developed across Asia over the last three to four millennia? If so, then the same should apply to Western teachings. I think such blanket statements refer more to academic understandings than appreciation of the essence of the religions. For this reason, I content that we can learn from the growing Western communication with the East through what is proving to be its most accessible form for Westerners, Buddhism. And it seems that this offers another perspective to understand the quest for sustainability. However, misconceptions of the essence of Buddhism, even among its adherents, pose an unnecessary barrier. The following discussion therefore introduces some teachings of the Buddha stripped of the cultural accretions that Buddhism attracted as it adapted to each new culture it affected.

One attempt to remove the obscuring cultural beliefs attached to Buddhism was propounded by the Thai monk Buddhadasa, a recent-day spiritual leader. He demonstrated a deep and idiosyncratic approach to Buddhism that allowed him to relate it to other religions. Thus the common intent of Christianity and Buddhism may be seen, even though it remains widely dismissed when literal comparisons are insisted on in both religions.[421] For example, Buddhadasa saw Jesus' words *do not suppose that I have come to abolish the law and the prophets; I did not come to abolish, but to complete*[422] as co-intentional with the Buddha's statement that *the Tathagata, the perfected one, appears in the world for the gain of the many, the welfare of the many, out of compassion for the world.*[423]

A case relevant to sustainability may be understood from consideration of the necessities (*paccaya*) according to nature (*dhammajati*) for food, clothing, shelter and medicine. These necessities that are recognized in the earliest of Buddhist scriptures are echoed in modern conception of rights of all persons to the basic needs for life. Buddhadasa saw this as providing a natural manner of understanding the intrinsic value of things (*vatthu*) - necessities have an intrinsic value while the value of all other things varies. An obvious example is the worthlessness of gold to a person if food and water are inadequate, but surplus food and water have no intrinsic value. This provides an inkling of the inherent value of natural morality, which includes balance, moderation and self-sufficiency. Expressed as the 'correct view' of the previous chapter, moral behaviour can be seen as contributing to a positive state of consciousness (*vinnana*) when we are content with the way things are (*prakati-sukha*). In colloquial terms, such calm happiness (Thai, *sangop-sukh*) is the antithesis of the modern usage of the Thai saying *eat well, live well* when the correct view would be something like *eat and live in a sufficient manner*. True morality (*siladhamma*) is therefore based on an ability to discriminate between the values of prices and natural

---

[421] Buddhadasa (1967)
[422] Matthew 5:17
[423] Mahasihanada Sutta Majjhima-Nikaya 12/37/46

values. Fitting into the natural values and their flows would thus be correct – and represent the basis of sustainability.

In contrasting a lifestyle that uses only what is necessary with modern lifestyles that depend on ever new things, moral behaviour (*siladhamma*) is a means of reducing greed. From this insight, Buddhadasa concluded that only a fraction of the input of conventional development projects is needed to obtain the natural output – that is the only output that could ever have been wisely expected. However, while worldly views do not value greed as a cause of problems in society, development agencies and governments routinely address only the symptoms of problems. To view un-sustainability as a problem without addressing greed is thus doomed to failure. The alternative is implied in Buddhadasa's *Dhammic Socialism* which warns against un-regulated capitalism and advocates self-sufficiency in agricultural production, ecologically balanced production of surpluses for moderate consumption by others, and wise use of and access to technology under righteous governance.

Another means of understanding essential Buddhist teachings is through the Indian concept of *ahimsa* (non-violence) as is discussed more fully in the next chapter. Usually translated as non-violence, *ahimsa* includes human attitudes to other beings, in contrast to common interpretations of Genesis and the writings of Aristotle, Augustine, Aquinas, and Descartes, which as discussed earlier, place animals at the service of humans. Commoditisation of animals in the West contrasts with Buddhist countries that, for example, had not until recently raised animals exclusively for slaughter.[424]

The co-incident Indian religion of Jainism has produced practical *ahimsa* approaches to technology that acknowledge the partiality of all truths and the consequent importance of

---

[424] Chapple (1993)

intention.[425] For example, in order to engage in the pharmaceutical industry, Jains must follow State legislation which requires the testing of products on animals before they are released to the public. The Jains' approach has been to rehabilitate such animals in shelters specifically maintained in association with their laboratories.[426] Is this an indication of a sustainable practice in terms of animal welfare? According to both Jain and Buddhist teachings, the answer lies in the intention not in the actions alone. If production of pharmaceuticals is motivated by compassion more than profit, then perhaps the industry has a component of sustainability. Such influences on the West are not new though they seem to be considered so in this era.

Both the non-violence and environmental ethics of modern Christianity have been traced to early Indian and Greek contact, possibly through Gnostic practices and Manicheanism, which itself had absorbed elements of the teachings of the Buddha, Zoroaster, Hermes, Plato and Jesus.[427] This has led to practices of vegetarianism in Christianity, sometimes with the ascription of souls to animals, although such practices and beliefs were periodically purged, especially by the Inquisition. Nevertheless, followers of St Francis maintained animal welfare aspects of the tradition, the Catholic Church retained abstentions from meat on Fridays, and vegetarian orders continue to today within both Catholic and Protestant traditions. Colonial contact with Asia expanded Western understanding, as is evident in Emerson's comments on the Vedas, the Laws of Manu, the Upanishads, Vishnu Purana, the Bhagavad-Gita and Confucius.

Emerson saw industrialized society as transient compared to the enduring insights of Asia. *The thought which these few hermits strove to proclaim by silence as well as by speech,*

---

[425] Chapple (1993) Page 45
[426] Tobias (1991) Page 31
[427] Chapple (1993) Pages 113-115

not only by what they did, but what they forbore to do, shall abide in beauty and strength, to reorganize themselves in nature, to invest themselves anew in another, perhaps higher endowed and happier, mixed clay than ours, in fuller union with the surrounding system.[428] With other contact through such means as the World Parliament of Religions of Chicago in 1893, Gandhi's use of the media, personal relationships and enhanced immigration, such concepts of *ahimsa* have proven at least intellectually appealing to such influential writers as Thoreau, a hero of today's environmentalists.[429]

Thoreau's ethical approach to the environment is used to postulate a continuum in land care in the USA. Thus he is linked to Leopold's statement of 1949 – *the land relation is still strictly economic, entailing privileges but not obligations. The extension of ethics to this third element in human environment is, if I read the evidence correctly, an evolutionary possibility and an ecological necessity.*[430] Building on this, Hargrove's conclusion of the importance of aesthetic appeal may now be seen in modern environmental sentiment.[431] Perhaps aesthetics too is part of a sustainable agriculture. But this slips into the common Western trap of dividing an insight into pieces in order to analyse it rationally. It is far from Buddhadasa's opening insights that moral action is acting in accord with nature which is sustainability, with intention determining one's relative success in acting sustainably.

Nevertheless, it may be that each culture finds its own way to morality. It does seem that the Western views above corrected Christian theology to a 'Franciscan worldview' in place of licensed exploitation. Attracting intellectual support if

---

428 Emerson (1903) Page 334

429 Thoreau (1999)

430 Leopold (1949) Pages 237-264

431 Hargrove (1989)

not specific changes within the Church, eight groups of continuing offensive teachings were elicited:[432]

- God transcends all, including nature
- Creation is depicted as a (male) act which confirms God's superiority over the created
- God created humans in His image separate from nature
- God gave humans dominion over nature
- God commanded humans to subdue nature and to multiply the human species
- The resulting power hierarchy of God, man, women, nature is inherently exploitive
- God is in humans but not nature which is a utilitarian product for both
- Aristotelian philosophy influences Christianity and compounds the view of nature as an instrument of man.[433]

In countering these views, Western scientific proclivities have led it to interpret them as an integrated nature, which is described in such terms as; [434]

- nature is an ecosystem of interdependent relationships,
- the life-force of each living being drives the system and may be likened to God in each being
- the system settles toward balance.[435]

Such thought is attractive to both Pantheists and Gaia adherents. Gaia theory postulates that rock-consuming bacteria created the conditions necessary for plant and animal life on earth thereby making air, rocks, and other seemingly lifeless matter biological rather than inert products. Likewise, without discussing the Gaia hypothesis in detail here, the universe may be conceived as a living system in which each

---

[432] White (1967)
[433] Callicott and Ames (1989)
[434] Capra (1975)
[435] Ingram (1990)

life-form's waste product is essential for others' life.[436] In many ways, this is similar to Jain concepts of 2,500 years ago.[437]

Is this relevant to sustainability? Yes, it is - for it illustrates how close Western society often comes to the essence of sustainability before shying away from it. Now the West seeks to draw concepts of sustainability from Buddhism, a group of profound teachings that eschew consideration of creation and conceive all things as interrelated in a natural flow. As discussed above, it is when we accord with that natural flow that we act sustainably - or in the language of Buddhadasa which we will discuss further, we accord with nature (Thai, *thammchat*), with the Dharma (Thai, *thamma*).

From this brief discussion of the West's encounter with Eastern religions, particularly Buddhism, we may elicit the following points:

- The similar teachings of Buddhism and Christianity have produced aberrant conclusions in both religions as cultural accretions distract from their spiritual essence.
- To live sustainably is to meet essential needs and to then develop wisdom sufficient to understand and live in accord with the dynamics of nature.
- Moral guidelines, both the great precepts and their derivatives, such as striving to minimize harm to any aspect of the natural order, may be understood as conducing to wisdom and hence sustainability.

The Indian spiritual consciousness that has produced and maintained valuable means of understanding Western teachings provides further insights relevant to sustainability, especially those concerning non-violence or *ahimsa*, which is discussed further in the following chapter.

---

436 Lovelock (1988)

437 Chapple (1993) Page 69

Chapter 21

Non-violence to the Environment: Active Sustainability?

What do you think, monks? If people carried off the grass, sticks, branches, and leaves in this Jeta Grove, or burned them, or did what they liked with them, would you think 'people are carrying us off or burning us or doing what they will with us'? – No, Enlightened One. Why not, monks? Because that is neither our self nor what belongs to our self – so too monks, whatever is not yours, abandon it; when you have abandoned it, that will lead to your welfare and happiness for a long time. What is it that is not yours? Material form is not yours...Feeling is not yours.... Perception is not yours.... Formations are not yours...Consciousness is not yours. Abandon it all and this will lead to your welfare and happiness for a long time.

Salayatanasamyutta, Sutta Nikaya

We can easily misunderstand most scriptures. The passage quoted above is far from an uncaring attitude to nature. It is addressing our attachment to materials and perceptions. It is supported by practical passages that deal with practices to become aware of our attachments to self and of the error of thinking that any thing is unaffected by and separate from any other thing. Such practical passages include ethical guidelines, a key one of which in Indian religions is not harming other things.

Eastern and Western religions including belief in science can be readily juxtaposed in terms of teachings and actions toward other components of nature. Indian religions include a pillar of non-violence (*ahimsa*), while Western religions have been interpreted to assign rights of dominion over nature and its exploitation. Yet the non-violence ethic in Buddhism exists in a creative tension between individual withdrawal from the self-oriented world and social engagement. Allegorically, the approach is assigned to the gods' delegation of the Buddha to share his awakening with all who continued to suffer the repetitive disappointments of life. Literal reading of texts on social issues were subsequently interpreted as support of forceful social engagement. But these actions missed the teachings about karma or consequences. Teachings of non-violence explain the decline of societies that ignore truth and morality. They may be better interpreted as macro-views of the wise more than as a history.[438] While literal understanding may be useful for governance, it does not reflect the intention of the teachings about non-violence.

Some analysts suggest that the non-violent ethic arose from rituals, in which the spilling of the collected blood of sacrificed animals was taboo. The ethic may have then evolved with the separation of the monastic community from society along with such matters as the dietary rules codified into the Buddhist *vinaya*. Whatever is the case, Buddhism seems to have adopted a pre-existing teaching of non-violence and to have used it to explain insights about the transcendence of the self. It may therefore be seen as an ethical guideline that considers non-violence as *self-assertion as much as self-denial*.[439]

Usually taught as the realization that all beings suffer, non-violence in Buddhism is an expression of acting lovingly (*metta*) towards all beings, which is interpreted as true compassion (*karuna*).[440] The Jain tradition, which arose contemporaneously with Buddhism, developed the non-

---

[438] Kraft (1992)
[439] Gomez (1992) Page 45
[440] Sangarakshita (1986) Pages 53-68

violent ethic (*ahimsa*) in a more pragmatic manner by classifying beings according to their apparent sentience.[441] And codification can lead to missing the spiritual insight that we are one with all beings. The Buddhist teaching has similarly been undervalued by unthinking obedience to such laws (*vinaya*) as the proscription of walking around in the wet season lest a small creature be trodden underfoot. As a means of making us aware of the value of all life as if it was ours, such rules have merit, but can easily become empty ritual.

And this is critical to understanding sustainability – following rules without understanding their intent is unlikely to produce lasting results. If one is genuine in the intention not to inflict violence on any component of nature, then one may approach sustainability. Of course, food cannot be produced without harming some micro-organisms and other life, but that is not the point. Literal interpretations do not help, either for understanding spiritual teachings or sustainability. The 'intent' of the teaching is to be aware of intentions as the primary determinant of outcomes. So the words that Hippocrates may have uttered apply to practical agricultural science – *first do no harm.*

We could rewrite the Hippocratic Oath for Agricultural Scientists in the following terms: *In appreciating the complex interrelationships between all things, I will respect all who have taught me this art and in the same spirit will impart knowledge of agricultural science to others. In diligently keeping abreast of advances, I will assist all who seek my ministrations, so long as others are not compromised thereby. I will consult specialists where needed and will follow methods beneficial to the environment. I will not administer any unnecessary environmental poisons and will respect all life. Except where essential, I will not interfere with nature nor carry out unnecessary research. May I enjoy life in the art and science of agriculture and the respect of my peers and society;*

[441] Chapple (1992) Pages 49-62

but if I violate this Oath, may the reverse be my lot. In such a conception, key words are 'essential' and 'necessary'.

Food is essential to life and its storage appears to be necessary under many circumstances. But neither of these factors is a licence for controlling access of other persons to food and its production, or to hoarding. Food production and its science are natural actions that fit into the natural flow, but intentions to gain unnecessary advantage at the expense of others upsets that natural flow. Ignorance of the natural flow in no way mitigates the suffering caused by 'good but ignorant intentions'. These teachings are clear in Buddhism and just as in other religions, they are supported by allegories.

The Jataka is a collection of allegories adapted to Buddhism from pre-existing tales that have separately come into Western culture as Aesop's Fables. They portray the Buddha's previous lives as moral principles. One example will suffice. *Once upon a time, a goat was led to a temple and was about to be sacrificed by the presiding Brahman. Suddenly the goat let out a laugh and then uttered a moaning cry. The Brahman, startled by this odd behaviour, asked the goat what was happening. The goat responded as follows: 'Sir, I have just remembered the history of what has led to this event. The reason I have laughed is that I realized that this is the last of 500 births that I have suffered as a goat; in my next life I will return again as a human. The reason I have cried is out of compassion for you. You see, 500 births ago I was a Brahman, leading a goat to the sacrifice. After killing the goat, I was condemned to 500 births as a goat. If you kill me, you will suffer the same fate'. The Brahman, visibly shaken, immediately released the goat, who trotted away. A few minutes later, lightning struck the goat and he was freed to become a human again. The Brahman likewise was spared, due to the goat's compassionate intervention.*[442]

Set in a time of animal sacrifice, the above tale made teachings of non-violence more intelligible to the masses. We

[442] Jataka Tale Eighteen, in Chapple (1992) Page 54

can argue that it is clearly not intended to be understood literally in a modern abattoir – but on the other hand, we do well to ask why in modern societies great effort is invested in separating killing from meat consumption. At the very least it places an undue burden for meat animals' welfare on the very few who do the dirty work. An intention to minimize the harm we do in everyday life would highlight this illogicality, which explains the association of vegetarianism with spiritual awareness. But vegetarianism is no more an ethic of Buddhism than it is of Christianity – it appears in the scriptures of both, as do passages that appear to condone meat consumption. And in both scriptures, it is the intention and situation of the act that is the key. Inflexibly following a personal 'law' is anathema to spiritual awareness. In societal terms, however, there is a benefit from general laws, which in large parts of the world extend so far beyond Western sensitivities that they are often dismissed as irrelevant, such as Indian attitudes to cattle. Yet one of the most reverent scenes I recall is of an Indian boy opening his wrapped rice and vegetables on the floor of a crowded railway station so that a cow that happened to be meandering through could eat. In such a case the 'law' about sacred cows may well have encouraged the intention.

Non-violence has been mentioned in law in India at least since the Buddhist king Asoka, who in one of his inscriptions recorded his conversion from warfare to non-violence and his requirement that the populace behave likewise. He further noted that *people can be induced to advance in the Dhamma by only two means, namely moral prescriptions and meditation. Of the two, moral prescriptions are the lesser, meditation the greater. The moral prescriptions I have promulgated include rules making certain animals inviolable, and I have established many other rules as well. But even in the case of abstention from injuring and killing living creatures, it is by meditation that people have made the greatest progress in the Dhamma.*[443] This legal enshrinement of animal protection has long been reflected in at least the letter of the law of Buddhist countries. In the case of Thailand limitation of killing of animals on the four religious days each month and a general government oversight of not allowing slaughter to exceed breeding rates was only modified in late twentieth century after extensive Western contact. Such a law is more intelligible within the Eastern concept of a righteous ruler (*raja*). *Raja,* while usually rendered as 'king', originally meant 'contentment' in terms of the leader's primary responsibility being the contentment of the community.[444]

Despite Asoka's clear communication of the role of ethical practices, Buddhist social actions are sometimes seen by Westerners as an inferior form of Christian charity programs.[445] This downplays the central aspect of intentionality in such sayings of the Buddha as, *a monk who has received ordination ought not intentionally to destroy life of any living being down to a worm or an ant.*[446] Non-violence (*ahimsa*) became widely known in the West from Gandhi's proclamations and actions and this is often a reference point for modern environmentalists.[447] As a practicing Hindu, Gandhi taught avoidance of destruction of life (*panatipata*) including *nature, plants and trees.*[448] It is but a small step from this to a definition of violence as *that which harms, debases, dehumanizes or brutalizes human beings, animals or the natural world.*[449] In this way, *ahimsa* is a component of sustainability.

The *ahimsa* concept was developed over centuries, finding mystical expression in the Mahabharata. The Yoga Sutra describes all things as *like a clear jewel, with unity among*

---

443 Swearer (1992) Pages 63-76
444 Swearer (1992)
445 Eller (1992) Pages 91-109
446 Mahavagga 1.78.4
447 Chapple (1993) Page 64
448 de Silva (1998) Page 114
449 Harris (1994) Page 2

grasper, grasping and grasped,[450] which may be understood to mean that, as all is self, all violence is against one's self.[451] As everything is in flux at all times, defining something as existing and not-existing[452] can only be a part of the truth when language depends on the limited perspectives of subject and object. From this reasoning, comes respect for the divergent opinions about *ahimsa* - including, obviously, modern views of non-violence to nature.

The most ancient scriptures, the Vedantic philosophies in fact contributed to an ecological worldview. Vedic hymns expressing an intimate relationship between elements of nature and humans.[453] Ayuvedic teachings describe the earth as mother and the individual person as the son, and observe that *I have settled upon [her], not suppressed, not slain, not wounded.*[454] More than modern pantheism, such insights were critical to the revelation of the nature of all things, which is flowing through the later Buddhism to which the West is being exposed.

The later Vedantic works and the Upanisads show the further development of human consciousness in similar terms of universal integrity. For example, in the Brhadaranyaka Upanishad:

As the tree of the forest, Just so, surely, is man.
His hairs are leaves, His skin the outer bark.
From his skin blood, sap from the bark flows forth
A stream, as from the tree when struck.
His pieces of flesh are under layers of wood.
The fibre is muscle-like, strong.
The bones are wood within. The marrow is made resembling pith.[455]

---

[450] Yoga Sutra 1.41
[451] Chapple (1993) Page 80
[452] Matilal (1981) Page 55
[453] DeNicolas (1976)
[454] Bloomfield (1897)
[455] Hume (1931)

Often revered only as poetry, such traditions have been retained in Hinduism and were exemplified in Gandhi's non-violent practices. And the Mahatma therefore refused to consider ethics separate from economics, thereby placing environmental, and in particular, social equity at its core.[456] The themes pervade the Indian contributions to our understanding of ourselves; for example Gandhi's ideal village. *My idea of village swaraj [self-rule] is that it is a complete republic, independent of its neighbours for its own vital wants, and yet inter-dependant for many others in which dependence is a necessity. Thus, every village's first concern will be to grow its own crops and cotton for its clothes.*[457] The amplified theme is similar when expressed in Western terms by Schumacher in his *Small is Beautiful*.[458] Nevertheless, actions spurred by such wonderful sentiments can be undermined when reduced to political acts.

It is easy to substitute the original spiritual development intent of ritual with political effect. Practical environmental expressions of Indian women embracing trees and the connection of trees by sacred threads to save them from mechanical clearing may well accord with local beliefs of wood spirits that protect human wellbeing. And existing practices in those situations may seem more sustainable than the alternative proposals. But when support for such actions becomes an internationally funded political act, it is difficult to see its spiritual purpose. As a consequence, it becomes more difficult to relate to sustainability. The divide is clear when the intentions of the various protestors are revealed to be as diverse as those of proselytizing politicians, evangelizing environmentalists, protectors of the people and economic expansionists.

While it may be true to suggest that commercial expansionism downgrades the moderating feminine traits that

---

[456] Gandhi (1959) Page 34
[457] Gandhi (1966) Page 43
[458] Schumacher (1973)

are more protective of nature,[459] its link to sustainability is weakened by the diverse, and thus potentially conflicting, motivations of its advocates. Similarly, the 'ordaining' of trees in Thailand by girding them with monk's saffron robes to garner local support against their being felled by developers separates the ritual from its original purpose.[460] Such actions are colourful and fun, and often effective, but appear to be peripheral to our quest for sustainability. The confusion they engender is spiritually distracting for the less sophisticated villagers who are too easily used as political pawns. Seen from this perspective, such actions can be an abuse of a power by external sympathizers, which itself is a form of violence.

So having said all that, what is the intent of Eastern thought with respect to environmental concern? It includes building on the insights of non-self, suffering and impermanence, a deep understanding of which leads to acting non-violently.[461] The concepts are hard to appreciate from an individualistic Western perspective, but may be explained in terms of differences in the meaning of the word 'community'. Development rhetoric may talk of constructing communities as interest groups, but in Buddhist terms it derives from local understandings in traditional lifestyles where *community for many native peoples is regarded as including other species, plant and animal, as well as environmental features and unseen ancestors and spirits. Community is lived and experienced as a whole system of interrelated types and species.*[462]

In modern Western language this might be expressed as living in the knowledge that our *day to day maintenance of our life support system is dependent on the functional interactions of countless interdependent biotic and physiochemical interactions. The inherent value of life is a core value in Buddhist codes and thus*

---

[459] Spanel (1988)
[460] Anon (1992)
[461] Chapple (1993) Page 64
[462] Hallifax (1990) Page 25

conforms to the notion of reciprocity and interdependence of a causal system.[463] The causal system and interconnectivity form the heart of Buddhist insights and are considered further in the following chapter.

Meanwhile, we may consider *ahimsa* in terms of interconnectivity. Buddhist abstentions on killing that derive from the *ahimsa* doctrine can be traced to an early belief that killing of any being produces negative karmic consequences. However, the intent of the underlying teachings was to both explain an insight of the wise and to engender compassion and benevolence as a spiritual development practice. Likewise, living simply is not meant to be an end in itself but a means of assisting spiritual growth.[464] Compromises accepted by Buddhism, such as in its expansion into China where it was initially perceived to undermine the integrity of Confucian family and economic values,[465] must similarly be considered if we are to understand the intent of their derived teachings. But it is of interest to note that the blend of Confucian familial duties and Buddhist personal spiritual development would seem to support self-sufficiency in such forms as family farms, which are one of the possible sources of sustainable agriculture.

Modern non-violent movements of socially active Buddhists face the dilemma of culturally modified teachings when they seek doctrinal support for their actions. By such considered testing of the teachings, we may comprehend such statements as those of Sivaraksa who claims that third-world farmer production of heroin, coca, coffee and tobacco could be conceived as preferable to cash cropping of rice or vegetables when the profits from food crops are reduced by international and national policies that favour the rich.[466] He seems to be

---

[463] de Silva (1990) Page 18
[464] Schmithausen (1991) Pages 38-53
[465] Ch'en (1952)
[466] Sivaraksa (1992) Pages 127-137

advocating the lesser of two forms of violence. Of course, our current high availability of food derives from both poor-country-agriculture and rich-country-industrial agriculture, and to increase the returns to poor countries may well reduce incentives for rich country producers and thus produce a food shortage. This discussion may offend sensitivities on both sides of the debate, for we prefer to see non-violence in physical and individualistic terms, just as we prefer to see sustainability as a means of self-preservation. We may summarize this chapter as follows:

- The ancient Indian ethic of non-violence reflects an integrated world-view, and guidelines developed from the ethic aim to assist the individual in spiritual development, even though they may be cast as society-wide rules.
- Rules concerning non-violence are often interpreted literally without awareness of their intent, which can produce anomalous situations in which greater violence is inflicted by following a rule, such as can occur in blind adherence to vegetarianism or naïve approaches to sustainable food production.
- Much of Western understanding of the non-violence ethic has been built on the political actions of Gandhi, which share the same insight that violence against anything is ultimately violence against oneself.

The interrelationships that make up all aspects of life and more are variously known as Causal Dependence, Conditionality, or Dependent Origination in Buddhism. Their relationship to sustainability is considered further in the following chapter.

Chapter 22

Sustained Change: The Conditions of Sustainability

Plus ça change
- plus c'est la même chose

If the concept of God is central to Christianity, then *Dependent Origination* or *Conditionality* (*paticcasamuppada*) is central to Buddhism. Put simply in the Buddha's words, it means – *if this is, that comes to be; from the arising of this, that arises; if this is not, that does not come to be; from the stopping of this, that is stopped.*[467] It explains Buddhist soteriology, suffering, psychological processes and more, to the extent that it also teaches – *he who sees the principle of conditionality sees the Truth. One who sees the Truth sees the Buddha.*[468] It is in this insight that the actions of sustainability become clear. However, to begin to understand Conditionality, we must remind ourselves of the insights about the self, the nature of existence, and the law of cause and effect.

The lack of any self (*anatta*) may be understood as life being the continuous coming together of diverse components. It is therefore our attachment to ideas, people and things that is the source of our discomforts when those things disappear or disappoint. As relationships with all things change when we understand these principles, so must our understanding of interactions with the natural environment. This leads to the conclusion that correct interaction of humans and the environment, which we might term sustainable action, is only

467 Majjhima-Nikaya II.32
468 Kulananda (1997) Page 7

possible by persons who are enlightened or truly wise. More practically, this means we begin to know about sustainability when we practice to develop wisdom.

The early scriptures also discuss this in the principle of *the three characteristics of existence (tilakkhana), which is a partial explanation of conditionality (paticca-samuppada).*[469] The three characteristics of existence are: the operation of the cause and effect principle, the durability of natural processes and their independence from a creator, external power or teacher. Amplifications in the scriptural Abhidhamma commentary divide natural law (*niyama*) into the five types of: physical inorganic matter (*utu-niyama*), which refers to weather, the seasons and the environment of human life; physical organic matter (*bija-niyama*), which refers to the continuation of species and genetic principles; order of act and result (*kamma-niyama*), which refers to human behaviour and its results; order of the norm (*dhamma-niyama*), which refers to interdependent causality; and psychic order (*citta-niyama*), which refers to the workings of the mind.[470]

When we conventionally discuss such matters as sustainable agriculture, we inevitably must cast it in terms of controlling such elements as the natural environment, genetic expressions, and general human behaviour. Yet insight of these matters shows that, for example, *if a person were the real owner of the [aggregates that make up life], then he could exercise control over these elements as he willed and not allow them to veer from a desired course or an ideal form that he would like them to maintain, such as desiring not to grow old nor ever become ill.*[471] As we do not control the elements, we cannot expect sustainability as long as we define it as something we want.

We might say that teachings about impermanence are naïve, and that it is obvious that all things change and

---

[469] Payutto (1995) Page 61
[470] Payutto (1995) Page 76, Footnote 16
[471] Payutto (1995) Page 63

degrade.[472] But the teachings indicate that the kind of change depends on its causal factors, which include our attempts to control (*yathakamma*). Control in turn leads to associated unintended changes as a result of our imperfect understanding of the interrelated operations of all things. That is, we act without sufficient wisdom. Once again, wisdom becomes the only hope for sustainability which relies on a myriad of factors that change continuously. Such wisdom is seen to arise from the calmness and neutrality described as the spiritually aware state of equanimity (*niramisasukha*).

Equanimity is displayed in living harmoniously with nature, which includes understanding and benefiting from natural law, without attachments to outcomes (*pannajivim jivitamahu settham*). To agricultural scientists this may sound similar to the continuous process of understanding nature. However, it varies insofar as modern science usually has an attachment to a commercial or other outcome that is superordinate to simply understanding the dynamics of nature as cause and effect.[473] In simple agricultural terms, certain causal factors are universally known, such as; *the causal factors concerning the germination and growth of a plant do not only indicate the existence of a seed, but also imply that the following elements must be present; soil, water, fertilizer, air, and proper air temperature. Each of these is a causal factor, and each is interrelated and does not have to exist in any orderly process according to a certain time and place.*[474]

It might then be thought that knowledge about ecological interactions is an insight into Conditionality. It is – but it is only one aspect. For example, the 'environment' itself cannot be examined as individual components unless countless interactions are also acknowledged. In the same

[472] Payutto (1995) Page 65
[473] Sayadaw (1969, 1981)
[474] Payutto (1995) Page 84

way, sustainability may be a self-made notion (*ahamkara*) to which our self is attached (*mamankara*).

Our creative and controlling ability has probably led us to create a creator to explain the cause of all things.[475] As tool-makers and creators, we logically conceive a higher order in terms of our own experience. But Conditionality has nothing to say about creator-gods beyond the way that our minds work, and in that context explains the problems caused by such conceptions. Whether the conception is of God or sustainability, the process is the same, and decisions made in ignorance of the mental and other conditions will inevitably cause unforeseen and unsatisfactory effects.

It is also in this way that teachings of impermanence (*annica*) show that the desire for sustainable agriculture or any other construct of 'self' is subject to the cycle of change and disappointment (*dukkha*). Without wisdom, such an outcome usually instils an even greater desire for the 'self' or the phantom of sustainability to continue, which leads to a cycle of disappointment until a glimpse of the higher truth is gained.

So far, we have discussed sustainability conventionally as an ability to control conditions sufficiently to lead to that state. But sustainability may also be seen as living harmoniously with nature. *Living with wisdom means being able to keep pace with changing conditions and knowing how to reap benefits from nature. Reaping benefits from nature relates to living in harmony with nature; to live in harmony with nature is to live freely; living freely means not succumbing to the power of craving and attachment; to live without attachments is to live wisely, or to know and relate to all things according to their causal factors.*[476]

This description of the process of Conditionality may be explained diagrammatically as cycles of existences, which are often portrayed as 'rebirths'. The cycle is driven by

---

[475] Payutto (1995) Page 91
[476] Payutto (1995) Page 95

ignorance, the origin of which is considered untraceable, thereby echoing the Judeo-Christian tradition's Fall of man.

The cycle of Conditionality may be expressed as in the following figure. **Ignorance** (*avijja*) of natural processes allows belief in a **mental formation** (*sankhara*) such as sustainability unless unwholesome tendencies (*kilesa*) that conflict with a holistic understanding are destroyed. This predisposition to the mental construct that sustainability outcomes can be achieved then leads to **consciousness** (*vinnana*) of the idea as apparently attainable **mind and body perception** (*nama-rupa*). Once consciousness of the idea arises, our **six sense bases** (*salayatana*) selectively inform the mind from external stimuli and **contact** (*phassa*) is made with the idea as if it is a reality, which then engenders **sensation** (*vedana*).

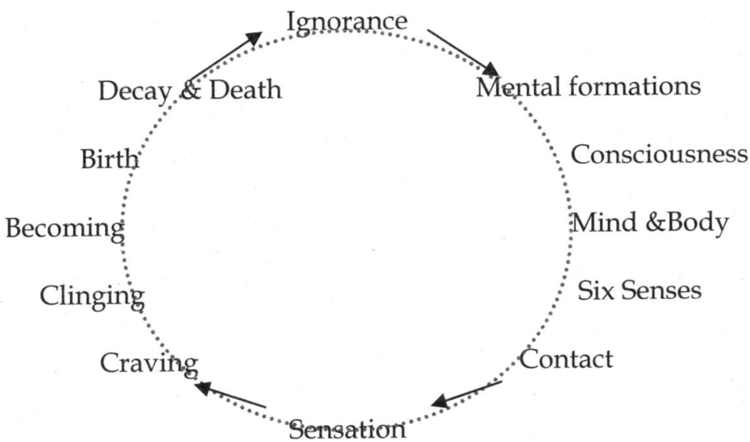

The Cycle of Conditionality

Continuing around the cycle, sensation comes **craving** (*tanha*) for the idea of sustainability, and for avoiding the unsatisfactoriness (*dukkha*) of its non attainment, which produces **clinging or attachment** (*upadana*) to the idea as one more false conception related to self (*attavadupadana*). So instilled with the self, the will (*cetana*) stimulates action (*kamma-bhava*), in what is termed **becoming** (*bhava*), which seeks to achieve the idea of sustainability through the actions set in train, and those actions then assume their own karmic effects. Karmic effects determine our reaction as the idea is **born** (*jati*) then **decays and dies** (*jara-marana*) and we see the attempt at sustainability is unachievable. Unless we review our mental state with wisdom, our state of disappointment (*dukkha*) becomes the essence of the **ignorance** (*avijja*) that first started the cycle. So the cycle is perpetuated until we become aware that *this whole heap of dukkha arises according to these factors.*[477]

Whether the cycle starts with ignorance (*avijja*) of natural processes or with craving (*tanha*) for sustainability, the result is the same - suffering. In craving for release from suffering (*dukkha*) another cycle based on ignorance is started. While some commentators claim that the cycle represents a human life span, it is intended to apply to all time frames and conditions down to the smallest thought-moment (Thai, *cittakhana*). The process may be further understood by considering the terms employed, which have multiple levels of meaning, for example:

- **Death** (*jara-marana*) includes separation from *having certain things and being a certain way,*[478] which is one common definition of sustainability.
- **Craving** (*tanha*) includes such sense pleasures (*kama-tanha*) as desiring the maintenance of something which one enjoys, and *in a deeper sense ... a desire for a self that can continue as it is or exist in a self determined manner,*[479] which applies to sustainability as a concept and its link to the desire for immortality.

---

[477] Payutto (1995) Page 105
[478] Payutto (1995) Page 117
[479] Payutto (1995) Page 122

- **Attachment** (*upadana*) includes *views, conclusions, theories and rules ... plans and strategies* which are perennially upset because, *when people have the life they want, there are undesirable things that will inevitably accompany it.* This produces such responses as *brainstorming, plotting, and searching – to find new ways to proceed according to their aims,*[480] which are constrained by thought being limited to such ideas as sustainability.
- **Ignorance** (*avijja*) confuses rational thought on a concept such as sustainability with behaviour *controlled by mental formations and predispositions, [which is] an utterly thoughtless driving force.*[481]

So attachment to 'sustainability' or most other things can only lead to disappointment. Apparent sustainability creates craving for more. Perhaps the unattainability of sustainability is its perverse attraction, fuelled by ignorance expressed as fear *that the self or pleasurable sensations (sukha-vedana) will disappear.*[482]

It is also relevant to sustainability that Conditionality applies in a positive manner and as such is the basis of Buddhist practices. This is correct view (*sammaditthi*). Early scriptures describe a series of *nadanas* that reverse the ignorant process by beginning with faith, joy, then *rapture, serenity, bliss, concentration, knowledge and vision of things as they really are, withdrawal, passionlessness, liberation, and Enlightenment.*[483] While it is hard to grok these terms without experiencing them, they indicate that craving, for example, can be used as a tool against themselves to interrupt the cycle. This suggests that environmentally aware actions performed for selfish reasons may yet help to increase awareness of ignorance (*avijja*) and so reduce the negative results from our actions

---

480 Payutto (1995) Page 123
481 Payutto (1995) Page 129
482 Payutto (1995) Page 131
483 Subuti (1995) Pages 66-68

(*karma*). As the early scriptures record – *this body was born of craving and depends on craving to sever craving*.[484] So the quest for sustainability itself could improve awareness of our self-serving and hence self-defeating actions.

The concepts summarized in this chapter are deep, complex and relate to all things; for the sake of our discussion on sustainability, we may condense it to the following points:

- The core of Buddhist insights, Conditionality integrates observations of cause and effect, the durability of natural processes, and the impermanent nature of all things, and in so doing highlights the only viable context for sustainability.
- The cyclical process of attaching ourselves to the idea of sustainability and being disappointed when it is not fully achieved is a classic description of suffering as self-induced and repetitive until wisdom allows a correct view.
- The same cyclical process operates for all aspects of everything, and in reverse is the source of Buddhist practice, as well as being the means of understanding how and when sustainability may exist.

Conditionality thus reveals the main conclusion that can be derived in our search for sustainability. From this point, which has been approached from many other paths in the chapters thus far, we may refine our understanding by considering other descriptions of Conditionality, such as *karma,* which forms the subject of the following chapter.

---

[484] Payutto (1995) Page 136

Chapter 23

Avoiding the Extremes: Karmic Sustainable Agriculture

Reflective virtue is simply an acquired clarity of the intellect, and moral virtue is constant warmth of heart kindled by that clarity. We should remember that of the human virtues none is more precious than discrimination ... For everything is an obstacle and nothing of use to a man who cannot distinguish the good from the bad and separate the bad from the good.

Marsilio Ficino

The description of Conditionality in the preceding chapter may have rational thinkers wondering if the insight is simply saying that there is potential good in all things and that most of our questioning is irrelevant. In one way that is what it is saying, as is explained in the Buddhist doctrine of Middle Truth (*majjhena dhammadesana*), which teaches detachment from extreme views. It has been scripturally summarized as *the Tathagata [enlightened one] does not attach to either of these two views and reveals the Dhamma in a middle way.*[485] With such insight, we see that too much philosophical investigation is pointless *because once someone sees dependent origination and understands that all things are linked in a process of cause and effect, these questions do not make much sense.*[486]

While it is anathema to rationality to challenge the scientific logic on which sustainable practices are based, Buddhist teachings indicate the limitations of rational thought

[485] Payutto (1995) Pages 136-145
[486] Payutto (1995) Page 143

without insight. We know rational thought is unreliable from personal experience and even in terms of what was once considered rational but is now held to be the opposite. Insightful understanding is gained through systematic and critical reflection (*yonisomanasikara*) that transcends belief (*saddha*) in a philosophy. Critical reflection leads to a clear distinction being seen between improper thought (*micchasankappa*) and its associations with satisfaction of sensual craving (*tanha*), selfish thoughts (*lobha*), resentment, malice and acting with ill-will (*dosa*).[487] Critical reflection on sustainability might thus identify such unattainable cravings as; an artificial stability, selfish maintenance of one's privileged position, resentment of the perceived causes of un-sustainability such as the effects of environmental and international exploitation. To avoid taking such extreme views, we might remind ourselves of the natural cycles that are working, whether they suit all of our sensitivities or not, for surely our seeking to protect ourselves from natural processes must produce some undesired karmic effect.

The doctrine of karma explains aspects of Conditionality and hence sustainability. False interpretations of karma relating to past lives, the will of the gods, or luck do not accord with such teachings. *However people sow their seeds determines their fruits: those who do good, receive good; those who do evil, receive evil.*[488] Thus karmic law applied to sustainability provides a simple explanation of the rising evidence of anthropogenic environmental decline and the potential benefit of personal and societal actions that aim to engage with other elements of nature in its natural flows.

Much of this may sound similar to Christian scriptures, and it is. But literal interpretation of the eschatological aspects of Christianity, which seem to pervade arguments about

---

[487] Payutto (1995) Page 232
[488] Payutto (1995) Page 153

unchecked exploitation of environmental advocates, are foreign to Buddhism - notwithstanding some errors that also arise from literalistic interpretations. The scriptures have the Buddha reiterate that he *will not speculate about the permanence or impermanence of this world.*[489] In so doing, they effectively define the scope for consideration of sustainability as the here and now, and this is the purpose of the moral guidelines of Buddhism, and all other traditions for that matter.

The moral code (*siladhamma*) of Buddhism aims to support spiritual growth, or understanding of the real nature of things, by according our behavioural and mental state to the development of wisdom (*panna*). Blind adherence to moral guidelines is seen as superficially beneficial but inconsistent with their real intent. In fact, attainment of wisdom produces moral conduct from the inherent correct action that flows from proper thought (*sammasankappa*), which with concentration (*samadhi*), leads to the equanimity (*upekkha*) of complete understanding. This state of wisdom is often described as an unbiased and unattached mind. It might even seem negative, for example when it considers statements clothed in sustainability rhetoric to be spurious because the motivations of those concerned are selfish. And to doubt the presence of the Emperor's new clothes is to offend everyone present.

Doubt of sustainability policies, research and actions sounds anti-scientific. But science offers tentative explanations not water-tight proofs and is ever redefining its conclusions. In the Buddhist conception, the fundamental truth of Conditionality and karma applies even when it contradicts current scientific belief. Thus the moral code of Buddhism is an expression of the cause and effect aspects of the natural processes used for personal training (*sikkhapada*) to avoid unproductive actions. Acting in accord with the law of karma contrasts with blind belief of science or theistic religion. It also

---

[489] Payutto (1995) Page 167

continually asks the individual to understand the truth behind a guideline as part of understanding the nature of things.[490] For to follow a material belief must produce angst when circumstances upset desires.

The inevitable and omnipresent angst of acting out of accord with nature[491] may be seen in forms as diverse as psychological problems and obesity. If we trick ourselves into a worldview that suits our comfort but assumes no change, we readily believe that change need not occur - that we can sustain things to suit ourselves. It is an understandable reaction; when too much is changing we grasp for something stable. And if we are honest, our search for something constant probably continues until we see the emptiness of our assumed unchanging self. This is one of the areas in which Buddhism and science in the form of psychology overlap.

We are easily deluded by our slow rate of bodily change through aging and by our recollection of dissociated events as a continuum to create our illusionary self. The self in this case is illusionary because regardless of what we think, it is changeable and will die. Our fixation on sustainability is best seen in our efforts to sustain our individual physical selves through medical interventions. Likewise, agricultural research is oriented to sustaining lifestyles, agricultural productivity, agricultural ecosystems, and profitability. As we have seen this approach to agricultural sustainability embodies an unacknowledged assumption that some hierarchy of rights exists in the world.

Part of the attraction of Buddhism to the West may be its lack of emphasis on hierarchies, which some modern Western thought interprets as a source of acts of dominance over nature. However, on further analysis such views seem naïve, as hierarchy is inherent to both Christian and Buddhist

---

490 Payutto (1995) Page 234-236, 247, 250

491 Claxton (1994)

traditions.[492] In Buddhism, hierarchies are assumed to explain the nebulous concept of enlightenment as a vertical hierarchy, and interdependency as a horizontal hierarchy. But this is different from the Judeo-Christian natural hierarchy of God-humans-animals-other. If we unthinkingly dismiss all vertical hierarchies as synonymous with domination and responsible for such problems as the 'ecological crisis', we may miss the critical Buddhist point that the vertical hierarchy of personal (and perhaps social) development is the common description of our movement towards wisdom. It is as critical as the horizontal path of recognizing interdependency. The latter is the more popular in Western forms of Buddhism and accords with modern scientific explanations of life, but it offers little more to our understanding of sustainability unless it includes the essential personal developmental element of wisdom.

In downgrading Buddhism to simplistic popular views of interrelatedness and as a solution to the worst in Western culture, the best of Buddhism may be omitted. Sponberg's *hierarchy of oppression* explains this as a rising degree of oppression on a vertical axis associated with a decreasing extent of interrelatedness on the horizontal.[493] He contrasts this with a *hierarchy of compassion* in which increased *evolution of consciousness* widens the degree of expressed interrelatedness. Such progress in consciousness confers an increasing ethical responsibility, which he refers to as *wisdom in action* – knowing the karmic effect of an action before it is taken.

When we recall that Buddhism teaches that all things, both material and immaterial, are entirely subject to the direction of causes, and interdependent,[494] we may begin a new form of communication based on the everyday scientific approach. Familiar concepts appear from these millennia-old scriptures in such forms as causal relationships of natural

---

[492] Sponberg (1997) Pages 352-376
[493] Sponberg (1997) Pages 365, 369
[494] Payutto(1993) Page 1

processes relating to physical objects (*utuniyama*), heredity (*bijaniyma*), works of the mind (*cittaniyama*), and human behaviour (*kammaniyma*), as well as a wider natural law governing the relationship and interdependence of all things (*dhammaniyama*). It is the penultimate of these that relates to karma as acts derived from intentions, which is the force that directs all societies and their values. These are not to be confused with social convention, customs, laws and beliefs; the essential difference is that under the law of karma, humans receive the fruits of their own actions in a natural process, whereas in social law, responsibility for one's actions is established by society. Of course, cosmic events that are unpredictable due to our limited knowledge still affect us, but this relates to the first type of karma, that of physical objects, and may not have any anthropogenic cause at all. This can lead to further confusion about karma when its association with intent (*cetana*) is forgotten.

As a partial explanation of Conditionality, an action can be divided into unskilful (*akusala*) and skilful (*kasula*) karma according to the relative presence of greed, hatred and delusion. Skilful karma includes connotations of general well-being, a clear and effective mind, wisdom and intelligence, and contentment. Its fruits are recollection (*sati*), love and goodwill (*metta*), non-greed (*alobha*), understanding of the way things are (*panna*), peacefulness (*passaddhi*), and joy at the good fortune of others (*mudita*). A shift between skilful and unskilful actions is so rapid that a highly trained mind is needed to distinguish between them. How could we ordinary folk possibly know the difference?

A sort of answer may be: *In this regard we can take a lesson from the situation of society in the present time. Human beings, aspiring to material wealth, holding a view that wealth of material possessions is the path to true happiness, have proceeded to throw their energies into material development. In the process they have wreaked destruction and untold damage on the environment.*

Now it has become apparent that many such actions were harmful. Even though society appears to be prosperous, humanity has created many new physical dangers, threatening the environment on a global scale.[495] Even though we might challenge the accuracy of such environmental observations, we are left with the guiding principle of intention as the most useful yardstick.

So from intention grow the fruits of karma experienced on mental, physical, life-experience and social levels. While most people focus on the level of life-experiences, it is the mental and physical levels that determine inner strength and external events. In the scriptural terms – *as the seed, so the fruit, who does good, receives good, who does bad, receives bad.*[496] This is further illustrated in homely terms, which incidentally include the law of heredity (*bijaniyama*); for example, *if tamarind is planted, you get tamarind; if grapes are planted, you get grapes; if lettuce is planted, you get lettuce. It does not speak at all in terms of social convention such as … if tamarind is planted, you get money, or planting lettuce will make you rich, which are different stages in the process.*[497]

The law of karma is a means of understanding Conditionality, yet even the fullness of the law of karma is beyond the comprehension of normal thought. It follows that when we seek to manipulate the law of karma to personal benefit, our intentions are unskilful. This unfathomability (*acinteyya*) of the law of karma is possibly a reason for its confusion with rebirth concepts. Here is a parallel with Western society's confused attitudes to sustainability, which include such elements as immortality (aka re-birth) and manipulation of karma (aka control of nature).

One clear expression of the potential application of skilful karma is the invitation to consider *the results on society and the quality of life even if one social value, that of materialism,*

---

[495] Payutto (1993) Page 27
[496] Payutto (1993) Page 41
[497] Payutto (1993)

were to change into an appreciation of skilful action and inner well-being as the foundations for true happiness.[498] In terms of ethics, understanding of the teachings about karma can improve everyday life if a person acts within natural law through skilful action informed by constant mindfulness.

If we consider the effect of our actions we begin to approach sustainability. But the message of Buddhism is of individual transformation, which contrasts with the common mass approaches of government, and such movements as 'Green Buddhism'. If we are to seek sustainability within Buddhism, then we may be easily confused. For *what Green Buddhism needs to explore more thoroughly is the Buddhist principle that meaningful change in our environmental practice can come about only as part of a more comprehensive program of developing higher states of meditative awareness, along with the increased ethical sensibility which this evolution of consciousness entails.*[499]

From our consideration of karma we may extract the following points:

- Naïve pursuit of sustainability can easily lead to dogmatic views that in fact seek to maintain an artificial situation, thereby providing a classic example of the action of the so-called law of karma when the effects fail to meet objectives.
- The universal law of cause and effect provides a basis for psychological exploration, which reveals an inverse relationship between acting with understanding of interrelatedness and oppression of nature and its components.
- From an understanding of the law of karma, the critical role of intention in determining an outcome makes skilful acts essential to sustainability without compromising the centrality of the transcendent message of Buddhism.

---

[498] Payutto (1993) Page 65
[499] Sponberg (1997) Page 374

A common conclusion can be derived from our multiple approaches in these chapters - sustainability can only be expected when our intentions are to act sustainably and are underpinned by wisdom. Trite as it may sound, the difference from our normal actions is easily appreciated when we realize that we usually use the laws of cause and effect to control nature out of self-interest. When we act in that socially endorsed manner, we expect that the results should be as foreseen without unfortunate contingencies. Enlightened self-interest, often represented as a pinnacle of Western civilization, is considered in the following chapter.

Chapter 24

Enlightened or Self-Interested? Sustainable Agriculture as Selfish

if you will to my thought revert,
that care for self is good for all,
if commune would to cash convert
all would balance and live life full!

Where this opening sentiment of the West prevails, its very prevalence is taken as proof of its truth. So it is logical for the West to think, 'if only all peoples would see it our way'. Intellectually it might seem that as more is learned about controlling nature, so skilful actions in terms of karma would become more common. Therefore, we might argue, if our intentions are noble and our knowledge advanced then we can expect to ever improve our comfort without side-effects. If we go a step further and assume that we are each responsible for ourselves and act nobly to others and other life forms, then we might even consider ourselves to be 'enlightened', in a civilized kind of way.

It seems that we comfortable people of the world do act as if we have some superior wisdom. We do in fact assume something of that sort when we suggest that each of our actions, even at societal level, can have an effect on others that may in turn have repercussions back onto us. One example is that of foreign aid: if, so rich nations argue, we do not help our poor neighbours, they will invade our shores and reduce our standard of living – therefore we assist them in our own self-interest. It is such 'enlightened self-interest' that is considered here in order to further clarify the meaning of the preceding two chapters.

Enlightened self-interest underlies attitudes to sustainability that encourage environmental care and social programs to avert negative impact on one's own wellbeing. It even extends to immortality through one's progeny, such as when it is piously argued that environmental care is on behalf of 'our children's children', as discussed in Chapter 2. We may also see a link between environmental care and economic growth or at least security. Viewed in this way, it is a selfish version of the Golden Rule; rather than *do unto others as you have others do unto you* it becomes *do sufficient good unto others to prevent them interrupting your own good*. Is this 'enlightened self-interest' in any way enlightened?

The adjectival 'enlightened' is clearly different from its use in Buddhism and similar religions. In fact, it even bears little resemblance to the 'Enlightenment' of the Renaissance. While this discussion may seem facile to some, or even polemical, antisocial, and 'un-American' to others, it is a means of linking some essential truths. It is a disturbing argument, for it highlights the futility of seeking sustainable solutions to anything based on 'enlightened self-interest', which includes most current 'sustainable' research and development programs.

Self-interest is a thick veil shielding reality from us today. Western society rewards individual effort in fields as diverse as science and sport. By individual effort the greater good is assured, it seems. But at the same time, it fails to notice the flaw in its logic when it claims the superiority of its system on the evidence of its own indicators of success, such as industrial output. Of course a materialistic society will produce more materials! But other societies may have [had] other value bases and rank the West's relative success differently. If the West only sees reality dimly as a result of this veiled glass as its own Christian tradition teaches, and if

Eastern definitions of the clear vision of reality are interpreted as enlightenment, then it can hardly be claimed that Western self-interest is enlightened. That is, in any sense except karmically shining a light on that self-interest. And that is how Western society and its individual attitudes are seen by wise observers.

The personal experience of enlightenment, about which much has been written - virtually all of it including this monograph in theoretical terms - can shed light on our understanding of sustainability. First, we must acknowledge that various attempts to classify the experience of enlightenment have foundered on the limitations of Western rational approaches for discussion of super-rational matters. Discussions of the psychology of nirvana confront us with the inherent bias of all investigators from their prior knowledge, opinions, ambitions, and intelligence.[500] Yet understanding of enlightenment exists in the West's own culture, such as in Blake's poetry: *He who binds to himself a joy/ Doth a winged life destroy/ But he who kisses joy as it flies/ Lives in eternity's sunrise.*[501]

Blake poetically captures the elements that the preceding chapters have summarized. Attachment is linked to pain, and detachment is linked to bliss (or eternity or immortality). These are the elements that will teach us sustainability - if we bind ourselves to the idea or even the utopia of 'our' sustainability, we will destroy that which we seek to sustain. But if we experience the joy of our state yet are detached from maintaining that happy state, we enjoy an eternal peace.

What does all this mean in every-day practical terms? We might, for a start, observe that the insightful awareness of enlightenment makes for a well-adjusted and effectual person.

---

[500] Johansson (1969)
[501] Blake (1996)

We might also observe that insightful persons are better able to judge when to act and when not to, thereby avoiding the common frenetic behaviour of modern society. This is also referred to as wisdom in the sense of Indian insights.

However, 'wisdom' is poorly served by its definitions in many dictionaries, as observed elsewhere. *Wisdom is even more difficult to define [than sustainability], not the least because the vast majority of us are not sufficiently wise enough to understand it! Most definitions dwell on the relative 'wise-ness' of persons, but the concept to which I refer as wisdom transcends ordinary discourse. We should therefore not focus on the common lexical presentation of the 'capacity of judging rightly in matters relating to life and conduct'. Subsidiary dictionary definitions provide an inkling of the meaning that I intend. For example, wisdom is used 'as one of the manifestations of the divine nature', as 'knowledge of a high or abstruse kind, enlightenment'; as 'wise discourse or teaching', and 'in the titles of two books of the Apocrypha'. I suppose that from such uses of 'wisdom' came its use for translation of the classical Indian concept that is ... an expression of insight about the nature of transcendent reality.*[502]

Such wisdom is undiscernible to most of us most of the time may be fleeting. It is not indicative of perennial infallibility or separate from individual idiosyncrasies. In fact, in the Buddhist tradition, the virtue of tolerance has long allowed for such variations; *the individualities of the Arahants [enlightened persons] were taken for granted and were respected and [individual's] self-assertions and defence mechanisms were to a certain extent considered as normal and permitted.* We also learn that *the idea of an Absolute Ideal is a European invention.*[503] From this perspective, we might consider that modern definitions of mental health share some approaches to enlightenment when they seek harmonious integration with values and purposes. If so, we can see sustainability from yet another perspective – as

---

[502] Falvey (2004) Pages 2-3

[503] Johansson (1969) Pages 135-137

a societal search for enlightenment, and as an ideal that cannot be intended to be an attainable practice or objective.

Approaches to enlightenment vary according to religions and traditions, but share essential elements. The 'evolving consciousness' described by the Christian Griffiths[504] and the 'higher evolution' analogy of the Buddhist Sangarakshita[505] share the biological evolutionary analogy in their explanations of the arising of a higher level of awareness resulting from personal commitment. Expressed in other terms this 'higher' path is that which treks back around the cycle of Conditionality discussed in the preceding chapters. So if common self-interest and enlightenment are opposing views, how do they relate to each other?

We often think that self-interest and enlightenment meet in persons who act out of individual self-interest within the limits defined by the group to which they belong. By contrast, a real 'individual' might be better seen as one who acts independently of group consciousness and takes responsibility for his own life and development. Once again, an interpretation in terms of sustainability suggests that the 'group' approach to sustainability inherently limits its success while a reflective and meditative individual may see the means of acting sustainably.

The emphasis on individual experience means that prescriptive approaches to 'right livelihood', to use a practical recommendation from Buddhist ethics, are likely to contain so many exceptions as to be misleading. This may not seem to be of concern when considering such obviously wrong livelihoods as trading in weapons, but may well be for many modern occupations. I do not think it is correct, although it is impressive to say one should *not be a pest exterminator, chemical farmer, genetic engineer, or exotic pet collector, [and] if we consider*

---

[504] Griffiths (1992)

[505] Sangharakshita (1994) Page 182

ads as intoxicants for products, advertising would also be a wrong livelihood.[506] Such extreme views would exclude many of the livelihoods of agriculture, mining, and medicine that contribute to current lifestyles. No, as we have already seen, the test is in the intent of the activity and the enterprise, not in an absolute rule.

Such matters are easier to comprehend from an historical perspective when we realize that we moderns strive for attainable goals in a manner foreign to views of self in earlier communal systems. We are self-reliant to a far greater extent, and see ourselves as *increasingly free from the bondage of nature; [mastering] natural forces to a degree unheard and undreamed of in previous history.*[507] The momentum of this rising view of self created the great revolutions of Europe and America. In Europe, the Reformation licensed individual relations with God free of church intermediaries and may well have been *the psychological preparation for the individualistic character of man's secular activities*, which were then moulded to socially useful functions by education.[508] And it is from this source that sprang the modern concept of sustainability among individuals separated from their culture's spiritual practices, myths and rituals that once engendered wisdom. And all that remains is a keen but unenlightened self-interest.

'Enlightened self-interest' may be better than blatant aggression to obtain whatever one craves, yet is clearly a long way short of the wisdom that allows sustainability to have meaning. We may summarize this discussion as follows:

- The concept of enlightened self-interest that underpins Western society can be readily revealed as selfish and historically biased towards material more than psychological development.

---

[506] Kaza (1999) Pages 54-75
[507] Fromm (1969) Page 127
[508] Fromm (1969) Page 313-4

- Self-interest obscures reality and thus precludes it being enlightened in the sense of Buddhism for example, which on the other hand could provide a vision of what sustainable actions in agriculture might be.
- Sustainability may be no more than an ideal in the same manner as full and continuous enlightenment, for which moral guidelines separate from wisdom cannot offer prescriptive rules, for sustainability or anything else for that matter.

Attempts at sustainability are thus hampered by a self-interested approach, which cannot be considered to be enlightened in any way except sophistically. This leads our discussion back to consideration of the relationship between real enlightenment and sustainable agriculture in Buddhist terms, as is related in the following chapter.

Chapter 25

Enlightening Agricultural Sustainability: Seeing More Clearly

Science cannot solve the ultimate mystery of nature because we are part of the mystery.

Plank

Science sheds light on details and occasionally embraces the intuition of wise practitioners to describe even more of the mystery. That experiential element of wise intuition is the focus of Buddhist and most other religious practices, if their cultural and institutional accretions are removed. So we now can amplify some key aspects of Buddhism that can shine more light on sustainability by considering the lives and works of two Thai practitioners. The choice of Thai sources simply reflects my own experience, but is also serendipitous as both Thai scholars have revisited the essence of Buddhist teachings within the less esoteric form of Buddhism, the Theravada tradition. Nevertheless, the following discussion of Buddhist insights could well have been prepared from other sources.

The Thai monk Buddhadasa says that Buddhism is *neither materialism [n]or mentalism, but is the correctness between the two or is both of them in the right proportions. The religion which can be taken as the best social science must not be a slave of materialism nor crazy about mental things*. This colloquial explanation is underpinned by his emphasis on the oneness of truth in all its forms, from the noble truths (*cattari ariya saccani*), to impermanence (*anicca*), not-self (*anatta*) or voidness (*sunnata*), conditionality (*idappacayata*), dependent co-origination (*paticcasamppada*), and 'thusness' (*tathata*).[509] In practical terms, this involves personal study and reflection on the Dharma, social responsibility in educating children in the Dharma, and mutual education in unselfishness. As discussed in the preceding chapters, these each relate to sustainability.

Where unsustainable practices cause suffering, be it in the form of reduced ability to produce essential food or just as a frustration about why we cannot achieve sustainability, the maxim ascribed to the Buddha is relevant – *I teach only 'dukkha' and the utter quenching of 'dukkha'*.[510] Reflection on these matters invites us to see that many concerns about sustainability arise from a misconception of reality, while others arise from an attachment to a standard of living.

Buddhadasa stressed spirituality in everyday life, emphasizing that the ideal of *nirvana* applied to the present rather than some future life.[511] In the state of *nirvana,* we can act in a sustainable manner because we act with wise compassion and loving-kindness. These 'fruits' of enlightenment are the same as the characteristics that one emulates in order to become conditioned to that state. One outcome of such conditioning is the acting in as sustainable a manner as is humanly possible as part of general morality. This is what he meant when he said *only genuine Buddhists can conserve Nature;*[512] speaking to a Buddhist audience, he emphasises *genuine* – we could replace *Buddhists* with *those practicing to understand reality*. The problem is ignorance of the operation of the nature and life. One means of understanding this is based on consideration of etymology.

In using the words for caring (*anarak*) about nature (*dhammajati*) as emanations of the pervasive human empathy that can transcend routine conservation projects, Buddhadasa

---

509 Santikaro (1996) Page 155
510 Santikaro (1996) Page 156
511 Santikaro (1996) Page 147-193
512 Santikaro (1996) Page 161

assumed that *dhammajati* includes everything connected to the Dharma (*dhamma*) and its origin (*jati*). Thus conservation of nature must mean an essential care for all things in the world in their natural conditions, separate from any benefit they might provide to us.[513] This etymological reasoning has parallels in the common Latin root *natura* for nature and birth.

From this realization, we may understand his fourfold explanation of nature: *Dhamma means Nature, which can be distinguished in four aspects: nature itself (sabhavadhamma), the Law of Nature (saccadhamma), the Duty of living things according to Natural Law (patipattodhamma), and the results that follow from performing duty according to Natural Law (pativedhadhamma).* As all four are known by the single Sanskrit word *Dharma* (Pali, *Dhamma*),[514] we like all other things including our creations, are part of nature, not separate or above it – a teaching that contrasts with the interpretations of Western cultural traditions, as discussed in the early chapters.[515]

Buddhadasa also examined other spiritual traditions, particularly Christianity, and found impressive similarities, which may be summarized in his words – *if we would like to have a God like they [other religions] do, we must take 'idappaccayata' (Conditionality) as God. It will be a more powerful God than any other ... God the Creator is nothing more than Conditionality, God the Destroyer is nothing more than Conditionality, God the Preserver is nothing more than Conditionality, Omnipresent God is nothing more than Conditionality.*[516] So sustainability requires perfect understanding of Conditionality, which itself is a definition of enlightenment. And the conditions that lead one toward enlightenment include conscious moral actions. Thus we may approach sustainability as an individual pursuit – not as an

---

513 Swearer (1994) Pages 1-26
514 APB (1990) Page 67
515 Santikaro (1996) Page 160
516 Santikaro (1996) Page 161-162

institutional ruling or as an extension message. In fact, in the form of a moral guideline (*siladhamma*), sustainability may be seen as the natural way of living.

From this viewpoint, sustainability is a description of the holistic (*kevala*) insights that understand it as the normal (*pakati*) condition, which includes avoidance of problems for others and oneself as a precondition for a moral (*sila*) society. However, Buddhadasa's own insights seem to indicate that seeking to avoid problems by mere observance of rules is morally egoistic and in itself productive of suffering (*dukkha*) as its not grounded in transcendent truth (*paramatha-dhamma*). Thus it is impractical to attempt sustainable agriculture in the absence of personal insight, particularly of Conditionality.

For about 2,200 years, Conditionality has often been interpreted as occurring across past, present and future lifetimes such that a past life's unskilful acts give rise to conditions in the present life, which in turn cause certain cravings or detachments in a future life. In correcting this Brahmanic influence, Buddhadasa showed the oneness of Conditionality and Dependent Arising (*idappaccayata* and *paticcasamuppada*).[517] This unifies other teachings and removes eternalist overtures, thereby according with the earliest teachings. It further indicates that karma is neither good nor bad, but simply the present reality (*sanditthiko*), which is not understood from intellectual consideration but from direct experience (*veditabbo vinnuhi*). That the word *paticca* means 'concerning', and *samuppada* means 'coming to be' also clarifies non-self (*anatta*) as the correct view that is usually obscured by our general preference for making a self 'come to be permanently'. And selfishness ignores the natural state as it seeks one's unnecessary desires ahead of others' needs.

With moral action (*siladhamma*) as the ordinary (*pakati*) state of the dynamic equilibrium of interdependence within

---

[517] Buddhadasa (1992)

which we come to understand our own relationship to nature (*dhammajati*), moderation, simplicity, empathy and mutual cooperation predominate. *If nature lacked this character we would all die. … [Farmers'] rice fields are planted for the benefit of wild animals who feed on it, as well as for their own consumption. They grow as much as they can to share with all forms of living beings.*[518] We may see the intent of some of Western culture's scriptures, such as *six years shall you sow your land and gather in its yield; but in the seventh you shall let it rest and lie fallow. Let the needy among you people eat of it, and what they leave let the wild beasts eat. You shall do the same with your vineyards and olive groves'.*[519] Expressed in his ecumenical terms, Buddhadasa said - *to be selfish is to rebel against God and Nature, to be the enemy of Nature and God. So let's end selfishness and reconcile ourselves with God, with the law of nature, with the law of idappaccayata. Then all the problems will disappear.*[520] His clear message is echoed in the more academic work of his contemporary, Payutto.

Payutto understands the teachings to suggest that an appropriate society is one in which individuals attain equal opportunity for self-development and wellbeing. Ethics are oriented to achievement of those ends.[521] While he does not isolate specific guidelines for sustainability, he implies it from self restraints from non-essential killing or doing harm (*panatipata*), or biasing conduct (*agati*) on the basis of personal preference (*chandagati*), fear (*bhayagati*), and foolishness (*mohagati*). We would associate only with true friends who shared the same vision (*mitta-patirupaka*).

In practical terms this includes such matters as scheduling release of water from dams equitably for industrial, agricultural or domestic uses, distribution of essential inputs, processing and collection of produce for

---

[518] Swearer (1991) Page 11-12
[519] Exodus 23:10-11
[520] Buddhadasa (1990) Page 12
[521] Payutto (2000) Pages 8-10

processing, and sharing of appropriate rewards.[522] Such peaceful coexistence is based on self-reliant conduct (Thai, *phuengtoneng*) achieved through the protective virtues (*nathakarana-dhamma*) of good conduct, learning of associations, approachability, communal contributions, truthful research, industriousness, contentment through moderation, mindfulness, and wise reasoning above emotional actions.[523] This would occur within a political environment consistent with the four principles of governance, originally defined as the king's support of the people (*raja-sangaha-vatthu*), which include:

- skilled promotion of agricultural technologies and policies (*sassamedha*).
- shrewd government administration (*purisamedha*) through policies and appointment of honest and capable officials who are provided with adequate social benefits.
- community support (*sammapasa*) through policies to support equity between all persons.
- impressive speech (*vajapeyya*) when advising, greeting and encouraging appropriate actions.

The specific mention of food production (Thai, *bamrungthanhaharn*) in many Buddhist scriptures indicates the pervasiveness of agriculture at the time. The attitude clearly differs from modern individualistic approaches to agriculture[524] where sustainability is most commonly sought. Yet, if sustainability is but an ideal or perception (*sanna*) rather than something that exists in itself, we must acknowledge that it too is subject to impermanence and change.[525] Our common reliance on perception often leads to distortions (*sanna vipallasa*), which has long been illustrated in Buddhism as a man mistaking a rope for a snake, or crows being scared by a

---

522 Payutto (2000) Pages 11-16
523 Payutto (2000) Pages 22-24
524 Payutto (2000) Page 30
525 Payutto (2001) Page 34

scarecrow.[526] In this vein, I wonder at times if we are actually distorting (*sannavipallasa*) natural environmental change into anthropomorphic apocalyptic environmental decline. Nevertheless, selfish ends certainly influence applications of science in agriculture.

Payutto takes the example of livestock feed additives, preservatives and flavour enhancers to illustrate unnatural interventions, which he attributes to a separation of science and business from ethics. Thus agriculture becomes *unrestrained production and consumption of goods with which to gratify the senses*, which leads to destruction and delusion.[527] This insight transcends simple analogies between the scientific method and the Buddhist Kalama Sutta,[528] which warns against belief in received knowledge, traditional practices, rumour, scriptures, guesswork, consistency with one's theory, seeming credibility, or faith in one's teacher. In fact the Kalama Sutta also cautions against reasoning in favour of personal experience for knowing the truth. Reliance on personal experience is not the same as the objective knowledge of science, and thus Buddhism is essentially disinterested in matters that engage science if they do not lead to wisdom. Just as the Buddha is cast in the scriptures as refusing to comment on specific scientific theories of his day, so we may miss the wisdom required for acting sustainably when we just focus on science as a solution.

Obviously science develops technologies that improve the human lot through equitable access and moderation, so with wisdom *sustainable development will surely become a reality.* [529] But a feeling that we lack something causes us to try to sustain the availability of consumables and profits and thus to bias science. This is our search for sustainability – a seeking of

---

[526] Payutto (2001) Page 48
[527] Payutto (1993) Page 14
[528] Kalama or Kesaputtiya Sutta A. I. 188 (60)
[529] Payutto (1993) Page 149

the security of permanence, and as we think we can control most matters in our lives, we feel we should be able to sustain our comfort. If the *unacknowledged repression of a spiritual character has played a significant role in the development of modernity*, it would seem that the science on which we rely may not be the primary source of sustainability.[530] From such wisdom, we may summarize this chapter as follows:

- In considering more of the unified insights of Buddhism, various perspectives emerge, all of which indicate that sustainable actions are natural and enlightened.
- As a natural means of behaving, sustainability is the operating of natural law, which is unrelated to suffering unless one is attached to an outcome.
- Science cannot produce a sustainable 'technology' as it suffers from conflicting objectives and ignores the reality of impermanence when it serves only material ends.

The insights of Buddhism show the constraints to sustainability as we usually conceive it. Other interpretations of these insights have stimulated apparently useful social development activities, some of which include sustainable agriculture projects, and it is to this 'engaged' Buddhism and some more scriptural references to agriculture that we now turn.

[530] Loy (1999) Pages 86-113

Chapter 26

Practical Buddhism: From Scripture to Sustainable Agriculture

This is a significant moment in the field of Buddhist studies ... the topic of Buddhism and ecology requires thoughtful and considered dialogue. [T]here is a hope that we can provide direction for future contributions of Buddhism to the problems of our contemporary society.[531]

If Buddhist teachings are oriented to personal development, some Buddhist social action programs appear to make an ill-fit. However, even if social reformation is not the intent of the early teachings, there remains a social action element in the Buddhist scriptures, such as links between good governance, which supports spiritual development and resultant sound social policies. And the ethical principle is to consciously act in a moral manner as part of one's personal development. Socially active and even political actions have long been undertaken by Buddhists, and today they tend to be called 'engaged Buddhists'. Their attempts at community sustainable agriculture may well inform us further about sustainability. We begin our consideration of engaged Buddhism by looking at the food production systems that sometimes confuse metaphor with ethical guideline.

The myths and metaphors that assist understanding of spiritual truths in Buddhism, as in Christianity, can be misunderstood in literal terms and for this reason are often dismissed. However, parallels between the religions are

[531] Lancaster (1997) Page 3

instructive in understanding our common natures with respect to the essentials of life, such as food. Just as the Fall of man in the Garden of Eden may be a description of our common dualistic thinking processes, which arises from the illusion of a separate self and its products such as vanity, lust and diversion, so may Buddhist myths reveal different meanings. According to one Buddhist story, food was once able to be gathered without effort, until we sought to control its production through agriculture, with higher levels of drudgery for sowing, harvesting and threshing.[532] The story is told in the Agganna Sutta,[533] which attributes the desire for control to the establishment of households and hoarding.

Such a myth reflects the friction between agrarian and pastoral societies, just as does the myth of Cain and Abel.[534] But it was probably used to illustrate the emerging social problems brought about by privately hoarding grain and owning land. In these ways, a contravention of natural cycles was related to acquisitiveness, greed and a selfish desire to control nature. Such an attitude, in fact, underlies most definitions of sustainability. Recognising this as our unredeemed state, another sutta discusses the role of the state in ameliorating poverty through social policies and property redistribution - failure to address poverty leads to social decline.

Other suttas indicate rising evil as a source of great awakening for the beginning of a new age, and in their structure make it clear that they are simply using that metaphor to teach about individual awakening.[535] As with Christianity, it is such texts as these that are interpreted literally by some to justify socially active Buddhism. Nevertheless, social action remains a useful practice for many

---

532 *c.f.* Genesis 3:19

533 Aggunna Sutta (2001)

534 Genesis 4:1-16

535 Hughes (1993) Pages 87-109

people to develop compassion, non-selfishness and other skilful actions.

The underlying Indian worldview of Buddhism had developed in a time when agriculture was the major economic activity, and this inevitably colours the teachings through parable and allegory - just as it does for Judaism. The Indian way of life assumed a hierarchy. Brahmans as the highest caste were responsible for rituals, *kshatriya* were responsible for administrative and military affairs, *sudras* were artisans and labourers, and *vaisyas* were merchants and businessmen. Association with profit-making was lower to reflect the spiritual aspirations of the society. The lowest cast, except for those outside the system such as foreigners, was for those engaged in agriculture and menial work. The Buddha, born into the *kshatriya* caste, treated all castes as equal and in so doing incidentally validated agriculture and commercial business. He used these widely in metaphors, such as in his 'fruits' of meditation to indicate the slow and progressive results of agriculture as distinct from the quick profits of trading which suit other metaphors. Of course, such scriptures are not describing agriculture but spiritual development.

In another instance, the Buddha is said to have described himself as a farmer whose tools of production were mindfulness (*sati*) and concentration (*samadhi*), which produced the harvest of enlightenment (*nibbana*). So Buddhism acknowledges agriculture as a fact of everyday life. While it has been observed that early Buddhist societies flourished while they focused on ethical agriculture and declined when they became too materialistic, the teachings are only using these as examples. So it is with non-agricultural environmental references. Nevertheless, the approaches remain useful in such areas as rural Thailand through the historic associations of monks with education, social action, wider communication, counselling and arbitration. But it is difficult to argue this as a basis for Western explorations of Buddhism.[536]

Modern Western fascination with the environment may offer a metaphorical link to Buddhist teachings. Such teachings apply to everything, not just 'the environment', as has been explained for lifestyles of two millennia ago.[537] *When kings are righteous, the ministers of kings are righteous. When ministers are righteous, Brahmans and householders are also righteous. The townsfolk and villagers are righteous. This being so, moon and sun go right in their course. This being so, constellations and stars do likewise; days and nights, months and fortnights, seasons and years go on their courses regularly; winds blow regularly and in due season … Rains fall seasonably, the crops ripen in due season … When crops ripen in due season, men who live on these crops are long-lived, well favoured, strong and free from sickness.* The spectre of climate change, potentially the major unknown in discussions about sustainable agriculture, menaces in such words insofar as some climate change is anthropogenic and continues through poor governance and flexible ethics.

The enduring nature of Buddhist ethical precepts (*sila*) requires explication to modern societies that are neither agrarian nor simple. Ethical principles are only flexible as regards purity of intention. Killing is unethical unless conducted in wisdom - wisdom of natural balances of animals, for example. Perhaps this means that when a rich nation exploits a poorer nation through biased trade arrangements and offers tokenistic aid in return it constitutes stealing, in a manner reminiscent of Blake -

Pity would be no more
If we did not make somebody Poor
And Mercy no more could be
If all were as happy as we.[538]

---

536 Prasety (1993) Pages 132-147
537 The Book of Gradual Sayings, Volume II. Page 85
538 Blake (1996)

I think this is what it does mean. But can logic be extended to espousing the *ecosphere of the planet* as a metaphor for the *sangha* – that is one's fellows on the spiritual path?[539] Of course it can, but it is metaphor not ethics. Yet some seem to benefit from the literalizing of metaphor, even as it distracts others from the self-transformation essence of Buddhism in favour of environmental revisionism.

Of course, an insightful person is not distracted and would clearly see the causes of environmental problems. In the case of nuclear waste, for example, an insightful person might well note the natural karmic consequences of creating long half-life radioactive material that affects future generations of many beings.[540] But it is the insight of consequences before an action is taken that has informed the development of ethical guidelines. In the absence of total insight, at least practice towards insight offers some self-awareness of one's intentions.

A more complex example of ethics and awareness at work is offered by apparent conflicts between relative goods. Is it better to move toward a more environmentally equitable human society with a small population living well on a renewable resource base, or to accept a high population living on an apparently declining resource base, or a few persons living well with large numbers barely surviving? I cannot suggest which is best, though some activists suggest the third scenario is the present and the first is desirable.[541] That seems fine if no-one is disadvantaged, but manipulation of persons to reduce their fertility can often be an abuse of power. Buddhism encourages moderation in consumption. It can be interpreted to advocate responsibility in reproduction if children cannot be well provided for without compromising their own and their parents' well-being. Of course, while the example indicates our flawed insight, it is simplistic; for

---

[539] Barnhill (1997) Pages 187-217
[540] Kraft (1997) Pages 270-290
[541] Gross (1997) Page 292-311

example, it appears that we know how to feed a population of 10 billion in a manner more sustainable than we do at present - but that is another subject. The example is sufficient to indicate that Buddhist ethics are of practical utility even with today's economic preoccupation.

Economic ethics in Buddhism can be found in such ancient scriptures as the Cakkravarti Sutta, which ascribes decline in society to the king's neglect of the poor. Poverty precludes practice of the Dharma when persons are distracted by hunger and other basic needs. The Mahasudaddana Sutta portrays the ideal society as providing food, clothing and transportation for all in need, while the Agganna Sutta criticizes the wealthy who cause hunger by increasing their holdings of rice fields.[542] A modern interpretation of the Vyagga and Kutadanta Suttas clarifies the essential difference between Buddhist and observed Western ethics of development; Buddhism recognizes the inherent value of environmental integrity and human livelihood, welfare and happiness, while Western ethics seem to subordinate these to outputs such as profit.[543] Applying these ethics in more than an individual application implies a society-wide approach to education based on Buddhist principles.

Education in Buddhism is based on self-reliance supported by association with those who provide good counsel, right instruction, encouragement and well-informed guidance towards enlightenment.[544] Self-sufficiency in this context can thus be seen to be an emotional commitment to others' wellbeing and to nature.[545] Such a vision is inconsistent with a selfish-materialistic education,[546] and incidentally also

---

[542] Tambiah (1968)
[543] Macy (1983) Page 45
[544] Payutto (1990) Pages 77-132
[545] Wasi (1999a)
[546] Wasi (1998) Pages 100-101

ignores the values of culture and tradition.[547] Attempts to return to such a self-sufficient system in Thailand, for example, led to a stronger community socio-economic base from which sound decision-making could have arisen[548] and where the principles of universal love and inter-relationship with nature might have been practised.[549]

The self-sufficient approach suggests that farmers are critical to overall societal wellbeing,[550] a circumstance which today is rare especially where low-input production systems pervade.[551] Defining 'development' in a different way to the usual 'economic development'[552] allows science to be integrated with Buddhist values while acknowledging the inter-relationships of spiritual, psycho-social, biological, and physical factors that relate to wellbeing.[553] From such discussion, we might elicit the following:

- Agricultural illustrations used in the Buddhist scriptures do not prescribe sustainable agriculture *per se*, but do indicate the counter-productiveness of hoarding, coveting and control of natural systems for selfish gain.
- Engaged Buddhism based on scriptural references to ideal social or agricultural practices is useful, unless it develops into beliefs to which others are encouraged to blindly subscribe.
- The central self-transcendent message of Buddhism is expressed at times in terms of sustainability, environmental integrity and self-sufficiency, and thereby indicates deficiencies in secular educational and development approaches.

---

[547] Wasi (1989) Pages 13-30
[548] Wasi (1998) Pages 41
[549] Wasi (1995) Pages 10-12
[550] Wasi (1999b)
[551] Wasi (1996) Page 26
[552] Wasi (1992) Pages 235-242
[553] Wasi (1994) Pages 11-30

Notwithstanding this reminder of the central intent of scriptures, we must acknowledge that the expansion of Buddhism into the West inevitably uses environmental examples and issues as a means of communication. This is fundamentally similar to the use of agricultural examples of two millennia ago. The following chapter therefore considers some of the environmental messages being embedded in modern Buddhism.

Chapter 27

Buddhism and the Environment: Wishful Ascription of Sustainability

> *Buddhism views humanity as an integral part of nature, so that when nature is defiled, people ultimately suffer. Negative consequences arise when cultures alienate themselves from nature, when people feel separate from and become aggressive towards natural systems. When we abuse ourselves, Buddhist ethics follow from this basic understanding. Only when we agree on this common ground can we save ourselves, let alone save the world.*[554]

While many of us were influenced by the first publication of Carson's *Silent Spring*,[555] technological manipulation of the environment has nevertheless accelerated over the subsequent four decades. Whether the current environmental crisis *threatens the very existence of all life-forms on the planet*[556] or not, there is a clear benefit of a sustainability ethic in agriculture, as there is in other human pursuits. However, we must be careful not to project desires for such sustainable actions onto the serendipitous concurrent introduction of Eastern thought into the West. For example, we recognize that *all traditional societies that have succeeded in managing resources well, over time, have done it in part through religious or ritual representation of resource management.*[557] But when we seek to isolate technologies from those societies for

---

[554] Kabilsingh (1990) Page 8
[555] Carson (1962)
[556] Tucker and Grim (1997) Page xv
[557] Anderson quoted by Tucker and Grim (1997) Page xviii

use in modern systems, we easily lose sight of natural rhythms. This is why science itself is often blamed for the 'crisis' – as if it is a component of the Judaism and Christianity that has exploited nature.[558]

Inappropriate action toward nature produces human suffering, according to the Buddhist scriptures.[559] Does this mean that environmental care is a central tenet of Buddhism because our ordinary conceptions of nature are limited by our senses?[560] I do not think so. The holistic view of Buddhism is inevitably vague when compared to a human-centred definitive approach, for Buddhism considers the psychological aspect of intent as more important than action itself.[561] Thus scriptures can teach that undesired occurrences inevitably accrue to whoever, for example, deliberately kills a deer without need and regret.[562]

Obviously, the environmental destruction in 'Buddhist' nations suggests that the intent of the teachings has been suppressed, though *it would be unwise to claim ... that Buddhism contains the intellectual and practical resources necessary to counteract the adverse effects of modernity*.[563] Do the vegetable metaphors for the Buddha in Japanese art and the emphasis on Buddha-nature in the environment in Chinese Buddhism offer a basis for developing a Western metaphor for the essential message of Buddhism as a useful form of eco-Buddhism?[564] Perhaps they do, but they are at best a teaching device for the intent of Buddhism, and at worst they can be a distraction from it.

Overall, the seeking of environmental teachings tends to subvert the scriptures' emphasis on the mental

---

[558] Lynn White (1967)
[559] Kabilsingh (1990) Pages 8-13
[560] Hayward (1990) Pages 64-74
[561] Keown (1995) Page 7
[562] Sutra for the Remembrance of Good Law ('Saddharmasmrtyupasthana Sutta')
[563] Harris (1997) Page 387
[564] Sueki (1994)

development that allows one to see through false realities, even though one form of false reality may be common conceptions of nature.[565] How has this misunderstanding come about? The first point that we must acknowledge is the pervasiveness of the Western value of environmentalism. Seeking environmentalism in the teachings of Buddhism often leads to Mahayana teachings, notwithstanding such apparent contradictions as, *nature is essentially possessive ... not natural ... until the hand of man ... shaped it.*[566] Such statements sit better with unreconstructed Western values!

Projection of Western ideals is also evident in the confusion that arises from attachment to English translations of Eastern scriptural references to nature. What is today often argued by Western Buddhists as evidence of a 'green' message in Buddhism includes reliance on such words as 'protect' or 'save'. In many instances these are translations of *raksha,* which in scriptural contexts originally meant 'observe' or 'practice' - words that refer to actions of personal development. In other words, 'observing nature' or 'practicing nature' referred to disciplined mental practice to understand the real nature of all things. It was not just the 'protecting' of nature as a rule, although at another level we may also interpret it as an ethical guideline as part of a reorienting mental process.

But religious practices are subject to cultural accretions that can mask their original intent. In the above case, the accretions may relate to the ancient Indian preoccupation with the separation between the body (*prakrti*) as a microcosm of the universe subject to decay while the eternal soul (*purusa*) is the principle of nature.[567] Such a tradition may conclude that eternal things are precious while changeable things such as nature have no intrinsic value. From such a tradition, Buddhism can be read to view forests as dangerous and

---

[565] Eckel (1997) Pages 328-349
[566] Ritchie (1971) Page 13
[567] Eckel (1997) Page 335

undesirable, while paradises are depicted as cultivated vistas landscaped to geomantic proportions. Grasping at such non-essential aspects of Buddhism by environmental advocates does not offer a link between sustainability and Buddhism. Essential teachings such as the tenet of non-self offer more in their encouragement of awareness of the integrity of nature.[568] To disaggregate humans from nature is to compare apples with oranges.

References to nature in Buddhist scriptures might also be understood as providing a holy environment for meditation and as an illustration of impermanence. This can be seen as a reason that we *have to take responsibility [our]selves for the harmony, the health, and the well being of the setting in which the quest for enlightenment takes place*.[569] Such a statement explains the concept of *Buddha-nature*[570] as a personal insight rather than an unrelated doctrine, as it has sometimes been interpreted in environmental arguments. So once more, sustainability may be seen as an outcome of insight rather than literal interpretation of ancient scriptures. It is easy to find environmentally responsible references in Buddhist scriptures, but it is also easy to find ambiguous references - just as both may be found in Christian scriptures. For example, modern Western interpretations of the egalitarian interrelatedness of all beings conflicts with the hierarchy of the natural world often assumed in Buddhist teachings.[571]

Understanding the conjoint use of metaphor and local examples can assist us to interpret such ancient views as forests being hostile, frightening, devoid of food and water, and the home of thieves, ferocious beasts and malevolent spirits.[572] These descriptions render the forest ideal for the

---

[568] Batchelor (1990)
[569] Eckel (1997) Page 346
[570] La Fleur (2000)
[571] Jorgenson (1989)
[572] *eg* Jataka 1: 332; 2:145, 335; 5:469-470

understanding of non-self, impermanence and compassion.[573] Tales of monks in forests abound from the Buddha's times until today. They seem to accord with the rising consciousness of animals as sentient beings, which are contrary to the thoughts of Aristotle and Descartes, express intentionality, emotion, and possibly forms of logic.[574] The Jataka stories - apocryphal stories of the Buddha's past lives that are probably derived from Sinhalese folk tales - simply illustrate the karmic consequences of wrong action towards animals and nature or the virtues of generosity, gratitude, non-injury and devotion. The same is true of the Jatakamala and the Mahayana Maharatmakuta Sutra.

The Jataka stories include ecological references in the contexts of rival water rights,[575] nature being protected through wisdom,[576] and the multiple habitats and integrity of nature.[577] Such stories need not suggest that Buddhism sees humans as reincarnated animals and trees, but it may indicate that Buddhism arose in a time of increasing agricultural and urban threat to forests. The stories strategically use beliefs and happenings of their time to describe how insight makes loving-kindness an automatic act of understanding interdependence. This is illustrated by the lines, *in the long course of samsara, there is not one among living beings with form who has not been mother, father, brother, sister, son, or daughter, or some other relative. Being connected with the process of taking birth, one is kin to all wild and domestic animals, birds, and beings born of the womb.*[578]

Environmental ethics must be practical. Buddhist ethics are also practical - although some sensible interpretation for

---

573 Kern (1943)
574 de Waal (1997)
575 Rukkhadhamma Jataka
576 Kusanjali Jataka
577 Vyaddha Jataka
578 Suzuki (1932)

modern scenarios may sometimes be needed.[579] As discussed in an earlier chapter, the field of environmental ethics has expanded to recognize global inter-relationships, future generations, non-human life forms, ecosystems and now sustainability. Some advocate improved management while others prefer radical changes in values and lifestyles.[580] This is one of the West's meeting points with Buddhism - a point where the self-evident truth of excessive consumption impacting negatively on nature becomes the metaphor for establishing harmonious conditions in one's own life. Such an approach could make sense of the environmentalists' mantra that Buddhism contains an essential environmental message.

Our conceptions of sustainability are supported by that term 'ecology', which Haeckel first used in 1866[581] and which has been further defined by Naess' 'deep-ecology'[582] with its associations of self-awareness. Such associations may be readily linked to Buddhist teachings. Through self-awareness we come to realize that we cause our own problems, and that we cannot solve all 'problems' by dualistic logic. Rather we require genuine compassion (*karuna*) and wisdom (*prajna*) to act skilfully.[583]

In the same way we may understand the Zen art and poetry teachings of Basho (1644-1694) as concerning self-awareness and interdependence. For example, *if you want to learn about the pine, become one with the pine; if you want to learn about the bamboo, go to the bamboo.*[584] This is close to the feelings expressed by many non-Buddhists engaged in agriculture - and such persons often practise more than talk about agricultural sustainability. This differs from the *spiritual*

---

579 de Silva (1998) Page 10
580 Dobson (1995) Page 2
581 Haeckel (1866)
582 Naess (1998)
583 de Silva (1998) Page 184
584 Yuasa (1966)

ambivalence[585] by which many environmentalists are characterized, which might equally be a rejection of codified religions. In any case, it remains fair comment that in some cases, environmentalists are simply repeating a rhetoric that is a *watered down remnant of Christianity with smatterings of Buddhism and Taoism thrown in to disguise the extreme blandness'*.[586]

The path between metaphor and meaning is paved with misunderstanding. The message of Buddhism about self-transformation tells us of the wisdom necessary for sustainable agriculture. However, the literal interpretation of metaphorical language must be expected to lead to conflicting conclusions, as was noted in earlier chapters where we discussed the intent of Christian teachings.

In recent times, the development of the Gaia theory represents a step toward alternative conceptions of reality. If the theory is interpreted in terms of homeostatic mechanisms that are not reliant on humans but on conditions that in turn affect subsequent events, then it is an example of Conditionality. However, the literal belief that all matter is part of a *single living entity, capable of manipulating the Earth's atmosphere to suit its overall needs and endowed with faculties and powers far beyond those of its constituent parts,*[587] restricts the definition of nature, for it omits the universe and our minds.

As the mind is the realm of Buddhism, it would presumably see the Gaia concept as just that - a concept, a general notion, an idea and therefore not a reality. The difference would seem to be that made between the algorithmic input our brains receive from our senses and the small part of actuality being perceived. However, in psychological terms, it may be a wider form of communication, of communion, with an organic whole of

---

585 Batchelor (1990) Pages 178-182
586 Batchelor (1990) Page 181
587 Lovelock (1982) Page 9

which we are part.[588] Such an explanation would seem to raise experiential knowing to a level on a par with other forms of knowing, and thus move a little closer to the Buddhist idea of mind.

The more common descriptions of the Gaia concept or model tend to follow either mechanistic forms[589] or apocalyptic beliefs. The latter viewpoint links the environmental decline induced by our *egocentric, self-referential, narcissistic* culture[590] to the overriding desire for continued material improvement,[591] belief-supported population growth[592] and the observation that societies that do not adopt this approach seem to be dominated by those that do.[593] Such fervour degrades the Gaia concept to be no different from any other bandwagon and of limited use to our discussion about sustainability – this is unfortunate as the concept is powerful and counters reductionist excesses. So once again we return to the conclusion that it is not a matter of just believing in scriptures such as tried in engaged Buddhist activities,[594] but of always being aware of the self-transcending intent of the teachings.

So when food is limiting, does it seem practical[595] to say, *do not suggest turning the earth over as creatures will be confounded*?[596] I don't think it does. Of course we may argue that today it could be practical, if we could correct food international trade inequities and use zero-tillage agricultural technology. But the scriptures did not foresee either of these approaches – they were saying that the priority was spiritual

---

588 Abram (1990) Pages 75-92
589 Volk (1997)
590 Hillman (1995) Page xxii
591 Singer (1993) Page 19
592 La Fleur (2000)
593 de Silva (1998) Page 35
594 Queen (1996)
595 Batchelor (1992) Pages 2-17
596 Sutta-vighanga

development and that in all our actions we should remain aware of our interrelationships with all other things. Similarly, the following myth of self-growing rice tells of our everyday grasping behaviour rather than some essential evil of agriculture. The myth relates how the naturally occurring rice that was able to be harvested whenever needed, was decimated by over-harvesting and hoarding of grain. This led to cultivated land assuming value and being greedily coveted for rice production with the concomitant outcome of craving for further land and theft of land and grain. Ultimately a ruler was required to control the increasingly complex society.[597] It is really a creation story of human behaviour that indicates our preference for the unreality of stability.

So once again, we return to reality. The reality is that our notion of a permanent self is challenged by the fact that it is dynamic and developmental, which means that the only sense of individuality of the self is as a karmic continuity. We are only the sum of our experience. Thus we might say that Buddhist 'environmental ethics' affirm our potential for higher consciousness, which impacts on nature through our practiced experience of compassion and altruism. It does seem that *turning to Buddhism simply as a traditional sanction for an already scientifically established ecological perspective on our problems adds little to what we already have.*[598] But it is also true to say that the mutually supportive developmental path of ethical conduct, meditation and wisdom produces the mindfulness and discrimination that allows the ultimate form of environmental care.

Of course, the compassionate expressions of Buddhism can easily be considered to be environmentally oriented as claimed by so called Green Buddhists. But in terms of Buddhist teachings, such a view can also be considered to be a

---

[597] Agganna Sutta (or Anguttara Nikaya)
[598] Saarmati (2001)

'near enemy' – a seemingly compatible view that in fact can divert from the essence of a teaching. In this case, Green Buddhism's general opposition to all forms of hierarchy can confuse Western students of the teachings, and in fact may be a Western projection onto selected teachings. Such a view of Buddhism narrows it to interrelatedness by ignoring the essential mental developmental component.[599] From such analysis, we may summarize this chapter as follows:

- In searching for means of expanding the message of anthropogenic environmental decline, Buddhism has been inappropriately interpreted to offer teachings that can literally inform sustainability.
- The teachings about natural rhythms form the basis of the holistic image of Buddhism and their true understanding offers a context for personal development that will inform sustainable agriculture.
- Western conceptions of the environmental message of Buddhism include modern Western projections, which limits understanding of the essence of Buddhism.

Notwithstanding this confusion over the environmental stance of Buddhism, we may appreciate elements of reality and wisdom from an awareness of our limited understanding of life. How this can develop further into a less worldly and more spiritual form of agriculture is considered in the following chapter.

---

[599] Saarmati (2000)

Chapter 28

In, Not Of, the World: The Spirit of Agriculture

Say of what folk by day and night
For ever doth merit grow?
In righteousness and virtuous might
What folk from earth to heaven go?
Planters of groves and fruitful trees,
And they who build causeway and dam
And wells construct and watering-shed
And to the homeless shelter give: –
Of such as these by day and night
For ever doth merit grow.

S Tr., I, 33.[600]

Now that we have seen that the primary contribution of Buddhism to sustainability is via its central message of self-transformation, just as it is for all spiritual traditions, we can briefly consider some specific interpretations of agriculture.

One reference to agriculture and its science (*sastra*) is attributed to Kautilya who was possibly a minister to King Asoka's grandfather. It classifies agriculture, cattle breeding and trade (*varta*) as a distinct and essential aspect of society along with various philosophies (*anvikshaki*) and government (*dandaniti*). *Varta* produced the grain, cattle, gold, forest products, and labour that underwrote the treasury, which enabled the King to govern (*dandaniti*) the kingdom and to support philosophy. By subtly raising agriculture from a low

---

[600] de Silva (1998) Page 125

caste activity to an essential component of a developing society in this way, the parity between agriculture and trade underpinned the emergence of Asian polities,[601] and this in turn has influenced the forms of Eastern religions that reach the West today. Thus their ethics take a practical form in support of their central intent.

In keeping with the society of its day, practical Buddhist ethics accepted some unintentional killing as a function of agriculture through such actions as ploughing and land clearing. Such pragmatism clarified what was, if interpreted literally, impractical ethical advice. Discourses of the Buddha address such injury to living beings in terms of awareness to minimize suffering to all things around the five themes of: sacrifices; warfare; meat consumption; suicide, and agriculture.[602] Yet all of these were accepted as part of life; so were such actions as ecological manipulation to create parks and shady arboretums (*aramaropa*), reforestation (*vanaropa*), and construction of irrigation, ponds and wells, all of which were considered to be meritorious deeds.

In a similar manner, the elements in Western society that we readily assume to be causes of environmental and social problems are not specifically condemned by Buddhism. Problems associated with property, capital and other aspects of economic behaviour are related to ignorance and attachment, not ownership *per se*.[603] Some persons can be just as attached to their views of environmental ethics as others may be to the profits of strip-mining. The same applies to competition, which is not condemned as such and, in fact, may be extolled in terms of excelling in one's pursuits. We should not be surprised that what some consider to be 'evils' are not seen in that light by Buddhism.

---

[601] Tambiah (1976) Page 27, 130
[602] de Silva (1998) Page 119
[603] Pryor (1991)

As we have seen earlier, it is intent or motivation that concerns Buddhism more than outcomes. This is why the scriptures appear vague and self-contradictory on specific environmental issues.[604] For example, the diverse economic references seem to advocate honest production and enjoyment of wealth without debt as much as living a full, interesting, creative, and enjoyable life in poverty - provided that one's lifestyle serves personal development.[605] It is such an understanding that informed some of the programs for rural development in Thailand after 1997[606] when self-reliance was promoted as a counter to excessive greed[607] and as a source of contentment.[608] In the same way, the Western tradition is wont to forget that it is the 'love of', (that is, *attachment to*) money, rather than money itself, that is the root of evil.[609]

When we assume that material wealth produces a good life, and ignore the inevitability of change of all things including ourselves, we must expect to be disappointed. Likewise a society structured on that assumption of selfishness, stability and sustained growth will exhibit inherent unhappiness. This is the Buddhist understanding of consumer societies.[610] Yet, the West is wont to mythologize its system as producing the greatest happiness. Buddhism clearly teaches that *from seeking comes the gain of profit, from gain comes discrimination, thence comes desire and lust, thence attachment, thence possessiveness, thence selfishness, thence hoarding, and from hoarding comes many evil and unwholesome things, such as the taking up of clubs and knives, quarrels, conflicts and disputes, recrimination, slander, and falsehood.*[611] Seeking to maintain

[604] Pryor (1990)
[605] de Silva (1998) Page 168-169
[606] Anon (1999)
[607] Adulyadej (1997)
[608] RPDB (1997) Pages 252-259
[609] I Timothy 6:10
[610] Bodhi (1999) Pages 43-53
[611] Mahanidana Sutta

wealth is often the most honest definition of modern 'sustainability'. Yet sustainability might alternatively be understood as a desire for continued 'security', which itself is one of the attributes - with happiness, peace, and freedom - of enlightenment or liberation from bondage.

This apparent confusion about sustainability expressed in religious versus other terms is readily clarified - for *a religion which points to the highest truth, to reality, will be in a position to unify with science. At that time science and religion will have reached another meeting point, their last one, where religion becomes science and science becomes religion, the division between the two gone forever.*[612] One means to move toward the meeting point is education, for *when education is out of balance, it only promotes our abilities to acquire material possessions and gratify the senses: ignoring our true potential, it fails to develop our ability to be happy.*[613] Such spiritually aware integration of science and education would facilitate a middle ground between urban restraint and nature conservation[614] and this forms the basis of recommendations for an alternative approach to agriculture[615] that is closer to 'Buddhist agriculture'.[616]

But can prescriptive forms of alternative agriculture really be deemed Buddhist? The choice of such alternative systems is broad, ranging from Thai self-sufficiency, to Japanese natural farming[617] and its refinements,[618] to Community Supported Agriculture, which links urban consumers to rural producers.[619] Unless the intention of not just the advocates of such systems, but also individual producers, is consistent with the teachings, then such

---

[612] Payutto (1993)
[613] Payutto (1994)
[614] van Willenswaard (1999) Pages 405-410
[615] Wasi (1998a)
[616] Wasi (1993) Pages 172-177
[617] Fukuoka (1978)
[618] Fukuoka (1987)
[619] Norberg-Hodge (1999) Pages 34-42

programs may be more environmental projects than Buddhist agriculture. Yet advocates continue to invoke scriptures to support their ideas.

The same applies to claims that new agricultural approaches are 'Judeo-Christian'. As we have already noted in earlier chapters, the Bible contains abundant agricultural examples: such as our recalcitrant attitude being described as *stiff-necked*[620] like working cattle unwilling to relax and lower their heads for yoking;[621] the story of a poor shepherd with only one ewe lamb that he treated as a child and which a rich man demanded to eat, told to symbolize David's murderous manipulation to have Uriah's wife,[622] and the exhaustive rules of Leviticus and Deuteronomy. But these too are metaphors and ethical guidelines for personal growth, and thus Judeo-Christian agriculture is the same as Buddhist agriculture insofar as it takes a liberated mind to understand what might be sustainable agriculture.

Yet agricultural metaphor, far from being a product of past millennia, lives on in rural Buddhist discourse. One recent example concerns the harnessing of a pair of draught buffaloes, one large and strong that provides power and the other small and clever that leads and obeys instructions. Representing technology, the stronger animal is contrasted with the wiser animal which represents true knowledge or Dharma to produce the conclusion that if modern technologies had been available during the periods of great spiritual advancement, they would have been applied to spiritual development. Buddhadasa extends his metaphor to observe that it is *incomprehensible that [even the best universities] do not teach about the purpose of life*.[623] On hundreds of other occasions, he also used traditional Thai proverbs, lullabies, folk stories

---

[620] *eg* Jeremiah 19:15, Acts 7:51
[621] Miles (1997)Page 136
[622] II Samuel 11&12
[623] Buddhadasa (2000) Page 22

and art to illustrate Buddhist teachings,[624] and to emphasise that the only appropriate science is that which leads to spiritual insight.[625]

To round off the subject, a final agricultural story from the Supreme Patriarch of Thailand is apposite. He relates the three duties of a farmer of ploughing, sowing, and irrigating to observance of the precepts, cultivation of mind, and development of wisdom. *A farmer does not have the power to say: 'may my crops spring up today, may the grains appear tomorrow, and may they ripen the next day' but when the time has come, the crops spring up, the grains appear, and they ripen. In the same way, [in] the undertaking of heightened virtue, the undertaking of heightened mind, the undertaking of heightened discernment, a Buddhist does not have the power or might to say: 'may my mind be released from the fermentations through lack of clinging today or tomorrow or the next day' but when the time has come, his mind is released from fermentations through lack of clinging.*[626]

So our search for agricultural sustainability has become the search for the purpose of life. It is a useful proxy in many ways; its metaphors pervade the writings of insightful persons, its everyday function in the times when scriptures were prepared meant that ethical guidelines contain much detail about agriculture, and its language is the language of religious practice. Such agricultural words as 'cultivation' (*bhavana*) continue to be employed even when we might otherwise use such terms as 'training' or 'practice' – an honour to agriculture[627] that can be overlooked by looking too hard for specific rules for sustainable agriculture. Surely environmental values that focus on integrity, stability, and beauty are complemented by Buddhist conceptions of change in *evolutionary, climatic, successional, seasonal, and stochastic* and

624 Gabaude (1994) Pages 27-59
625 Santha-Anand (1994) Pages 61-70
626 Nyanasamvara (2001)
627 Falvey (2001)

other forms.[628] Our common longing for permanence, which is the source of the quest for sustainable agriculture, is thus revealed as usually nothing more than a craving. However, in a few cases it may be as a product of great personal insight. In any case, as an ethical ideal, sustainable agriculture remains a useful practice.

So from this chapter, we may conclude that:

- Attachment to the letter of ethical laws, to prescriptive agricultural ethics, or to the imposition of personal ideals onto others is not consistent with the development of a sustainable agriculture.
- Acting to sustain productivity, income or anything ignores the inevitability of change as revealed by both science and religious insight, which can be the same thing when dynamism is seen in sustainable agriculture.
- Agricultural examples, agricultural ethical guidelines and agricultural language pervade scriptures to explain the cultivation of higher consciousness, and this is turn may allow the understanding necessary to sustain agriculture.

Oriental texts, including the Buddhist scriptures tend to wander widely back and forth across a subject to illustrate lessons from different perspectives. For we in the West, this can seem unstructured and repetitive. At the same time, it is ironic that the West's current social eschewing of hierarchy is not reflected in its preferred mode of discourse, which remains a hierarchical logic. The foregoing chapters might be seen as a short Asian journey across the subject to obtain glimpses of the reality of sustainability from different perspectives. That journey is mapped more concisely in the concluding chapter.

---

[628] Callicott (1997) Page 63

Chapter 29

Tying the Thread:
The Sutra of Sustainability

There, on the mountain and the sky,
On all the tragic scene they stare,
One asks for mournful melodies:
Accomplished fingers begin to play.
Their eyes mid many wrinkles, their eyes,
Their ancient, glittering eyes are gay.
Yeats

Our meandering search for sustainability has eclectically selected philosophical elements from diverse opinions in order to divine a common thread. By tracing that thread of the suture (Sanskrit, *sutra*) that binds the body of wisdom, we connect our search for sustainable agriculture back to some dubious motives. We also understand the intent behind common justifications for environmental activism. Now, as we follow that *sutra* through each of the chapters, we may describe how Yeats' old men could be happy amidst the tragedies that we bring upon ourselves.

We began by considering sustainability as a proxy for beliefs in immortality or rebirth and revealed a fundamental fear of impermanence. This fear of impermanence explains the resilient misinterpretation of eternity common to all religions and which sanctions the popular belief in sustainability as a virtue for science to investigate. While pure science studies the processes of nature and recognizes cycles of life and decay, the more narrowly conceived technological or applied sciences are oriented to forestalling change - and death, the ultimate change. This occurs even within cultures that consider themselves secular.

The so-called secular culture of the West retains its Judeo-Christian origins in the form of an assumed licence to dominate other elements of nature. By misinterpreting the creation myths and thus denying their lessons about the interconnectivity of all things (in God or heaven in this conception), time, space and quantity have been emphasized as the elements of sustainability. At the same time, a fascination with a deterministic model of an inconceivably complex nature has fostered neglect of what has long been termed the divine essence in all of us. Neglect of this essential part of natural well-being produces the indeterminate angst prevalent in modern lifestyles. Yet, ironically, it is these modern lifestyles that demand sustainability at the same time as continued growth. Its criticality is then shown by invoking an environmental apocalypse that mimics literalistic eschatology.

Literal readings of the Old Testament easily elicit concepts of apparent relevance to agricultural sustainability in such forms as stewardship, equitable land distribution and simple close-to-the-earth lifestyles. But such interpretations must be tempered by the emerging human self-consciousness expressed in the historical narrative of everyday activities, which include agriculture in metaphors for personal development. From this perspective, references to sustainable agriculture separate from personal self-transformation are not readily apparent in the Christian scriptures.

The spiritual insights of Christianity revel in the changeability of life as an expression of universal divine immanence. This is the Christian version of the theme recurring through this book – *sustainable agriculture exists within dynamic change*. For those who have exchanged literal definitions of God in favour of an experience of an immanent spirit, sustainable agriculture may be seen as working within and respecting the spirit. For others, such awareness might also be expressed as God's hand in managing natural law. However, those who adhere to a paternal controlling God inevitably arrive at a belief in human-controlled sustainable agriculture that varies little from rationalist conceptions.

Since the Reformation, the West's trust in rationality has diluted the impact of insights of human unity with an unseen order. It is thought that statements about sustainability are rational and whoever challenges rational thought is regarded as today's heretic. Yet what is called a rational search for sustainability may simply be self-will, which when it ignores the natural order, can produce little of lasting benefit. Such unsustainable outcomes, as argued through this book, are an example of what Christian language calls sin - failing to accord with the natural unity, or if you like, reality.

Insights of human unity with all things have commonly described a divine indwelling that permeates all nature and its flows, which confirm that agriculture can only be sustainable when practised within those flows. But technology seems to disrespect those flows as it seeks to understand only that part of them that technologists or their masters wish to manipulate in response to a societal fear of change. This could explain the myths and symbols that refer to psychological health being dependent on knowledge of the integrity of all things - the first premise of sustainability, and an insight that pervades all great religions.

All religions centre more on our common potential for higher forms of consciousness than everyday experience, although each religion has its own cultural and institutional accretions that mask this essential message. The Old Testament reveals the emergence of awareness of human consciousness and realization of the ideal of balance between physical, psychological and spiritual development. Within this context, scriptural references to agriculture as a punishment or source of human misery may be understood as rhetorical in the same manner as the pain of childbirth, and as emphasizing the karmic interconnections between all things from thoughts to actions.

In a more advanced form, prophetic insights speak of reality in terms that Western rational interpretations can only appreciate as a dim reflection. So it is possible that the emphasis on sustainable agriculture is a misguided response to mythological and allegorical messages about our oneness with all things. But it may simply be the desire to maintain things the way they are – a kind of agricultural salvation from the vicissitudes of change. Or an emotive reading of immanence may erroneously project spirituality into all things, which would suggest that ecological and agricultural sustainability can be a form of belief-based pantheism, or of God, without the option to further develop one's consciousness.

Pantheism is not some new (or ancient, as the church once taught) form of devil worship. In seeing the divine in all nature, pantheism can encourage an interrelated conception of life that fosters positive environmental attitudes which can be a step towards sustainable agriculture. Its apparent appeal mirrors literal interpretation of maternal metaphors intended to explain our separation from our natural state when we ignorantly exert our will. However, pantheism's ambivalence toward personal transcendence denies it a role in rational and spiritual development, both of which appear to be critical to sustainability.

Such developmental elements may occur through philosophy and science as much as through religion in the ethical evolution of improved human relationships with each other and nature. Just as the West now conceives of greater rights for slaves and animals than in the past, so it is beginning to appreciate the wider rights of nature as a secular ethic of sustainability. When there is an overemphasis on mechanistic agricultural approaches to profit from nature, animals are treated unethically and even the basic elements of sustainable agriculture are neglected. The Western societal response expressed in the emergence of rights may also be seen as part of an evolving self-consciousness, although this may be diluted in secular ecological ethics.

Unguided by ethical guidelines or insight, anthropocentric attitudes to sustainable agriculture have produced a utilitarian basis for evaluating sustainability. This secular ecological understanding has produced a well-meaning but largely impractical theory of ecologically sustainable agriculture that ignores the fundamental ecological manipulations that define agriculture. Logic requires that we seek sustainable agriculture within agricultural-ecology rather than outside it in the first instance, yet the separation of food production from consumption allows misinformed urban ecological sensitivities to restrain the contributions of religion, philosophy and science. Such an inherently artificial separation of modern secular and traditional religious ethics is unlikely to persist in such a pragmatic field as agriculture.

The convergence of values from religious, secular and foreign traditions has already produced an eco-theology that recognizes agriculture as an essential activity that should be practised within natural flows. At the same time, the church has reacted against pantheism by renewing emphasis on the ethic of stewardship. But in fact the role of religion has declined to the extent that interpretation of societal feelings is now the preserve of secular philosophy. And as neither have produced modern ethics specific to agriculture, it probably is considered mainly within the philosophical field of the 'liberation' of nature or its elements.

While secular philosophy interprets moral trends in the rights of animals and nature that are compromised by utilitarian valuations, the problem-solving approach of sustainable agriculture research maintains a reactive mentality. For example, animal agriculture continues to assume that animals do not feel pain in the same manner as humans. Likewise, applying a commercial model to all agriculture when the priority of most of the world's farmers is subsistence alienates the West from natural morality and thereby negates attempts at sustainable agriculture. Nevertheless, in exposing the unstated assumption of continued economic development, secular philosophy could in fact relate sustainable agriculture to a wider morality, which one might expect to see reflected in approaches to global economic development.

It is at the juncture of sustainability within dynamic natural flows and sustainability within economic development (sustainable development) that global economic discussions reveal their peculiarly Western orientation. Emphasis on environmental issues in poor countries is a relevant example in that growth once assumed to be limited by the availability of natural resources is now delimited by ever new technologies. By constant innovation the modern developed-country technicians assume that they can sustain whatever they or society desires. Some observers see this as a pragmatic component of consumer capitalism that the West is extending across the globe – a perpetuation of old Western ways in a new world, albeit clothed in new rhetoric in aid and NGO programs. With faith in innovation as a tenet of sustainability, new agricultural technologies are continuously demanded to maintain economic growth, and the West evangelizes less-developed countries with this model. Thus 'sustainability' becomes the servant of economic growth and existing global hierarchies, which adds little to the understanding of real sustainability.

The motivation to sustain Western lifestyles ahead of others, even though the contrary is claimed in international rhetoric, is masked by polarized views around competing definitions of economic and ecological sustainability that must be unified if they are to contribute to agricultural sustainability. Driven by Western secular thought, sustainability in less-developed-countries increasingly means policies to stabilize population and reduce poverty through economic growth. This can be warranted, except where economic growth assumes cash cropping and ignores subsistence farming to the detriment of both the local community and the local environment. As the West learns this lesson, it is becoming aware of useful traditions that it has long abandoned.

Traditional agriculture retains attitudes to nature that contrast with those of intensive modern agriculture and suggests that sustainable agriculture is indissolubly joined to attitudes to nature. Traditions of peasant farmers may not always offer obvious productive advantages, but they do recognize agriculture as a modified landscape integrated into the wider natural environment. This may be as close to a sustainable agricultural ecosystem as we have approached so far, which suggests that the West must rediscover its lost spiritual dimension of agricultural sustainability. Recombining the spiritual and intellectual dimensions allows greater acceptance of insights that link overall health to holistic agricultural practices. Such linkages between agriculture and spirituality in the West, where they have survived or where they are mindfully recreated, appear similar to those of peasant agriculture.

When one realizes that the diverse interactions of a small-holder agricultural system in a less-developed country cannot be rigidly defined, it becomes clear that agriculture cannot be defined by technology alone. The intrinsic interrelatedness of nature described through religious metaphor is foreign to scientific discussions. Yet even though the West has excised spiritual aspects from agricultural science, elements of a unified understanding survive in smaller scale agriculture and gardening where participation in nature is valued above level of output, as is common in Eastern traditions.

As the West (re)discovers Oriental worldviews, it respects the re-linking etymological intent of 'religion' and the continuing evolution of knowledge of ourselves. This stimulates reconsideration of ancient scriptures across various traditions. Buddhism seems compatible with scientific discourse through its insistence on cause and effect in natural flows. Taoism may also fit this description. However, while the apparent compatibility seems to benefit science, it may not benefit the transcendental message of such religions if it limits them to scientific logic. In Buddhist language, science in such a guise may be seen as a 'near enemy' of the teachings as it draws the mind away from its potential to transcend such limited views. At a more practical level though, religious insights provide guidelines for our everyday interactions with nature.

Moral guidelines drawn from insightful observations of natural flows provide, for example, an indication of the origin of the concept of Christian sin as acting out of accord with that flow. When the exotic cultural accretions of Buddhism are stripped away, its expansion in Western societies can be seen to assist in explaining both the West's tradition and its infatuation with sustainability. The essential teachings of Buddhism and Christianity concur that living a sustainable lifestyle leads to contentment, once basic needs are met. It is wisdom to live in accord with the dynamics of nature. Ethical guidelines can then be seen to be self-training actions that lead to wisdom and hence sustainability.

Differing interpretations of one ancient Indian ethic, non-violence, illustrate how anomalous outcomes may occur when awareness of intent is replaced by blind observance of a rule. This may be seen in fundamentalist vegetarianism, and even in approaches to sustainable food production that eschew scientific knowledge. On the other hand, the core of Buddhist insights - Conditionality - shows the self-induced suffering caused by attaching ourselves to such an idea as sustainability. The reverse path is Buddhist practice - acting ethically to condition oneself to instil an automatically ethical behaviour, which when coupled with mental development, produces a mind less congested with fears, frustration and anxieties. Even at a cursory level, it can be seen that Conditionality reveals sustainability to be reliant on wisdom, which itself is the product of mental discipline uninfluenced by unhealthy desire. Thus intentions and actions determine each subsequent event in ways that ignorant understanding cannot conceive.

The naïve pursuit of sustainability leads to a desire to maintain artificial situations which inevitably prove unsustainable, thereby providing a classic example of the operation of karma. This universal law of cause and effect also provides a basis for psychological exploration, which reveals the inverse relationship between acting with understanding of universal interrelatedness and oppression of nature. Before action, intention ultimately determines the skilfulness of an act with respect to sustainability and all other matters. Thus sustainable agriculture can only be expected when the intention is to act in a sustainable manner and it is underpinned by wisdom that accepts the incomprehensible interactions of nature. However, just as a little knowledge is said to be dangerous, so perhaps is a little wisdom, if such a thing exists. The law of cause and effect is often interpreted to mean that nature may be manipulated for some 'noble' purpose with the expectation of good results without unfortunate contingencies - and secular worldviews misguidedly call this enlightened self-interest.

Enlightened self-interest appears to be inherently selfish and to derive from an historical bias towards material more than psychological development, which obscures the view of reality and thereby precludes it from being seen as enlightened in any real sense. Thus self-interest leads away from what sustainability may really mean, for sustainability is an ideal in the same manner that full and continuous enlightenment is an ideal. Just as moral guidelines are not prescriptions for enlightenment in themselves but form one part of a process of development, so prescriptions for sustainability may be futile without the concurrent development of the wisdom needed to live in nature.

Practising agriculture in a sustainable manner is probably both a natural and an enlightened action. As a natural action, sustainability is the operating of natural law, which is the realm devoid of the suffering engendered by attachment. Attempts to produce sustainable agriculture from applied science tend to ignore the conflicts of multiple objectives, the reality of impermanence and the fundamental search for that elusive something that is forever unattainable through material means. Thus, the insights of Buddhism and other traditions, clarify the constraints to sustainability. But it would be foolish to accept popular versions of Buddhist agriculture as the solution to this dilemma, just as it would be foolish to accept literal Christian interpretations.

Notwithstanding implied environmental messages in Buddhism, prescriptions for sustainable agriculture *per se* do not exist. Some claim that unsustainable agriculture is shown to be derived from hoarding, coveting and control of natural systems for personal gain. Socially-engaged Buddhism indeed uses such scriptural references for its justifications. But when these develop into proselytized beliefs, the central self-transcendence message is lost, even though at times that message is expressed in the language of sustainability, environmental integrity and self-sufficiency. As Buddhism expands in the environmentally-conscious West, environmental issues inevitably become a means of communicating deep truths in a manner similar to the use of agricultural examples of two millennia ago - but the medium should not be mistaken for the message.

When enlisted in support of the message of anthropogenic environmental decline and portrayed as offering direct ethics that can inform sustainability, Buddhism is but a political device. In truth, the holistic image of Buddhism does offer a context for the personal development that can inform sustainable agriculture, just as does Christianity. But attachment to the letter of ethical laws or prescriptive agricultural ethics, no matter how apparently righteous, is still attachment - and attachment is the source of suffering according to Buddhism. Acting to sustain productivity or income ignores the inevitability of change as revealed by both science and religious insight, and that very dynamic is what has ever been known to wise persons as the context of sustainable agriculture. Agricultural examples and language pervade scriptures written in agricultural ages and it is from these that we are all taught to 'cultivate' the higher consciousness that produces the 'fruits' of wisdom, and one of those 'fruits' can be sustainable agriculture.

So, from this search for sustainable agriculture from the perspectives of both science and spirituality, we return to that place from whence we set out and *know it for the first time*, as Elliott also found. We can dispense with the motherhood cachet of sustainability and accept it for what it is - our best attempts at maintaining something that suits us. But we can also accept it as an indicator that agriculture is never likely to be sustainable unless we work within the natural order. By definition modern agriculture does not work within the natural order - it seeks to maintain an isolated unnatural order. Until recently, humans only had the power to modify the natural order minutely and usually in ways that would be pushed aside by nature if they overstepped the mark. Nevertheless, the deserts of sand, erosion and salt to which most cultures have contributed across the millennia testify to either our ignorance of, or our lack of concern for, agricultural sustainability.

We would do well to also recall the coincident expansion of deserts and social ills across history. Rising Western psychological ills seem coincident with a fervent interest in sustainability, which means that we cannot just conclude that this is a simple projection of an inner need for stability and permanence onto science and religion. It is much better to address the cause of these yearnings directly rather than through such proxies as 'sustainability' or additional consumption of goods and services, or blind belief in a salvific image.

Our yearnings are older than agriculture, although they come to us via the writings and traditions of our tribal agricultural forebears. We once expressed these as yearnings for a lost Eden, a lost Golden Age, a forgotten jewel, a return to our father's house from our life eating swine swill, and a hundred other stories from various religions. Today we search for the same things expressed in different words surrounding 'sustainability', and we waste much effort and resources on seeking a solution when all the time it has been within us, in the very place from which we set out. Expressed in biological and spiritual terms, sustainability can only be realized by insight into our place in the natural order.

Sustainability Sutra

Is it God's grace or nature's law, or logic founded in folk lore
or heretics who's more is less, or prophets' cries in wilderness,
that protects from the pain of change, and anguish over rights estranged –
from slaves, to cows, to plants and nature – when we focus on our future?

East exhibits what we once knew, lost spirit limits what we do,
for sin is shunning nature's flow, whether in heaven or here below;
the only God – the god of karma, told in apocalyptic drama,
which self-interest has exposed, as vain – to all virtue full opposed –
unless it conceives all as related, that selfish urge is ne'er sated.

Yet we life's lessons still ignore, and promulgate our new-found 'law':
'if you will to my thought revert, that care for self is good for all
if commune would to cash convert, all would balance and live life full.'

Oh! medium mistook for message, renders religion to the savage,
casts wise fruits before blind swine, thus we define land, food, as 'mine';
our lust for lost golden ages, long lectured to us by the sages,
whose saga of forgotten jewel, ever within for our renewal,
if we from sleep but ourselves rouse, and so return to our father's house,
then we re-turn and see the rhyme, know whence we left for the first time.

So as we trace that thread sublime, and let the bells of wisdom chime,
spurn literal texts where truth's neglected, where no-one knows all is connected,
thus surely see all is sustained, for only then is Eden gained –
but he who'll but on sense rely, reality doth yet deny.

References

Abraham, Ralph (1990) Orphism: The Ancient Roots of Green Buddhism. Chapter in Badiner, Alan (editor) Dharma Gaia: A Harvest of Essays in Buddhism and Ecology. Parallax Press, Berkeley. Pp265.

Abraham, Ralph (1994) Chaos Gaia Eros: A Chaos Pioneer Uncovers the Three Great Streams of History. Harper, San Francisco.

Abram, David (1990) Perceptual Implications of Gaia. Chapter in Badiner, Alan (editor) Dharma Gaia: A Harvest of Essays in Buddhism and Ecology. Parallax Press, Berkeley. Pp265.

Adulyadej, Bhumibol (1997) Royal Address Given at the Ceremony of the Northern Agricultural Seminar, Northern Agricultural Office, Chiang Mai. February 26.

Aggunna Sutta (2001) Sahitya Akademi. Pp101.

ANCARD (1993) Economy, Ecology and Spirituality: Toward a Theory and Practice of Sustainability. Asian NGO Coalition for Agrarian Reform and Rural Development, Manila. Pp28.

Anderson, Jock (1993) Does Sustainability Make Economic Sense. International Symposium on Trade and the Environment, sponsored by the Minnesota bar Association, Minneapolis Convention Center, Minnesota, November 10-12.

Anon (1992) Interview with Phra Prachak. Seeds of Peace 8:1:32.

Anon (1999) Self-Reliance Theory of Development. Journal Siam Society 87:xiii-xv.

APB (1990) Dhamma Propagation Book of Meanings. Alliance for the Propagation of Buddhism, Bangkok.

Aristotle (1916) Politics. Dent, London.

ATSE (2001) http://www.atse.org.au

Baer, Richard Jr. (1966) Land Misuse: A Theological Concern. The Christian Century 83:1240

Bailey, Liberty Hyde (1915) The Holy Earth. New York.

Baillie, John (1933) And the Life Everlasting. Scribners. Pp293.

Balfour, E.B. (1975) The Living Soil and the Haughley Experiment. Faber and Faber, London. Pp383.

Banks, Robert and Stevens, Paul (1997) The Complete Book of Everyday Christianity: An A-Z Guide to Following Christ in Every Aspect of Life. Intervarsity Press, Illinois.

Barnhill, David (1997) Great Earth Sangha: Gary Snyder's View of Nature as Community. Chapter in Tucker, Mary and Williams, Duncan (editors) Buddhism and Ecology: The Interconnection of Dharma and Deeds. Harvard University Center for the Study of World Religions, Cambridge, Massachusetts. Pp467.

Batchelor, Martine (1992) Even the Stones Smile: Selections from the Scriptures. Chapter in Batchelor, Martine and Brown, Kerry (editors) Buddhism and Ecology. Cassell Publishers Ltd, London.

Batchelor, Stephen (1990) Buddhist Economics Reconsidered. Chapter in Badiner, Alan (editor) Dharma Gaia: A Harvest of Essays in Buddhism and Ecology. Parallax Press, Berkeley. Pp265.

Bentham, Jeremy (1843) Principles of Penal Law. In the Works of Jeremy Bentham. Bowring, John (editor) Volume 1; London. Pp562.

Berger, P. (1914) William Blake: Poet and Mystic. Chapman and Hall, London. Pp420.

Berry, Wendell (1981) The Gift of Good Land: Further Essays, Cultural and Agricultural. Northpoint Press, San Francisco. Pp115.

Blake, William (1996) Auguries of Innocence. Selected Poems. Penguin. Pp242.

Bliss, Perry (1926) The Heart of Emerson's Journals. Houghton Mifflin. Pp368.

Bloomfield, Maurice (1897) Hymns of the Atharva-Veda. Oxford University Press, Oxford.

Bodhi, Bhikkhu (1999) A Buddhist Social Ethic for the Next Century. In Socially Engaged Buddhism for the New Millennium: Essays in Honor of Venerable Phra Dhammapitaka (Bhikkhu P.A. Payutto) on his 60th Birthday Anniversary. The Sathirakoses-Nagapradipa Foundation and The Foundation for Children, Bangkok. Pp536pp.

Bookchin, Murray (1986) Post-Scarcity Anarchism. Consortuim Books. Pp315.

Bookchin, Murray (1988) Toward an Ecological Society. Black Rose Books. Pp315.

Braden, J. and Kolstad, C. (1991) Measuring the Demand for Environmental Quality. Elsvier, New York.

Brockway, Allan (1973) A Theology of a Natural World. Engage/Social Action 23:37.

Brundtland, Gro Harlem (1987) Our Common Future. World Commission on Environment and Development. Oxford University Press, United Kingdom.

Brundtland, Gro Harlem (1987) What is Sustainable Development. In Towards Sustainable Development. The Panos Institute, London. Pages viii-x.

Bucke, R.M. (1905) Cosmic Consciousness. Innes, Philadelphia.

Buddhadasa, Bhikkhu (1967) Christianity and Buddhism: Sinclair Thompson Memorial Lecture 5th Series. Thailand Theological Seminary, Chiang Mai. Pp 125.

Buddhadasa, Bhikkhu (1990) Talk sponsored by the Komol Kimthong Foundation, Pacarayasarn Journal, and the International Network of Engaged Buddhists.

Buddhadasa, Bhikkhu (1992) Paticcasamuppada: Practical Dependent Origination. Suan Mokkhabalarama, Chaiya, Thailand. Pp116.

Buddhadasa, Bhikkhu (2000) Life Should be Harnessed by Two Buffaloes. Translated by James Ratana Nantho Bhikkhu. The Buddhadasa Foundation, Suan Mokkhabalarama, Chaiya, Thailand. Pp 45.

Callenbach, Ernest (1975) Ecotopia: The Novel of Your Future. Bantam, New York..

Callicott, Baird (1988) In Defence of the Land Ethic: Essays in Environmental Philosophy. Albany, New York.

Callicott, Baird (1997) Earth's Insights: A Survey of Ecological Ethics from Mediterranean Basin to Australian Outback. University of California Press, Berkeley. Pp292.

Callicott, J. (1990) The Metaphysical Transition in Farming: From the Newtonian-mechanical to the Eltonian-ecological. Journal of Agricultural Ethics 3(1):36-49.

Callicott, J. Baird and Ames, Roger (1989) Introduction: The Asian Traditions as a Conceptual Resource for Environmental Philosophy. Chapter in Nature in Asian Traditions and Thought: Essays in Environmental Philosophy. State University of New York Press, Albany.

Campbell, B. (1989) Seven Ages of Conservation Farming in Victoria. Australian Journal of Soil and Water Conservation 2(3):12.

Capra, Fritjof (1975) The Tao of Physics. Shambhala Publications, Boulder, Colorado.

Capra, Fritjof (1996) The Web of Life: A New Synthesis of Mind and Matter. Harper Collins, London.

Carson, Rachael (1962) Silent Spring. Houghton Mifflin and Riverside Press, Cambridge, Mass. Pp368.

CGIAR (1987) The CGIAR: Facing Tomorrow's Challenges Today. The Consultative Group on International Agricultural Research, Washington DC. Pp16.

CGIAR (1997) Annual Report of the Consultative Group on International Agricultural Research, Washington DC.

Ch'en, Kenneth (1952) Anti-Buddhist Propaganda During the Nan-Ch'ao. Harvard Journal of Asiatic Studies 15:1&2:166-192.

Chandler, M. and Holliday, S. (1990). Wisdom in a Postapocalyptic Age. Chapter in Sternberg, R. (editor) Wisdom: Its Nature, Origins, and Development. Cambridge University Press, New York.

Chapple, Christopher Key (1992) Non-Violence to Animals in Buddhism and Jainism. Chapter in Kraft, Kenneth (editor) Inner Peace, World Peace: Essays on Buddhism and Non-Violence. State University of New York Press, Albany.

Chapple, Christopher Key (1993) Non-violence to Animals, Earth, and Self in Asian Traditions. State University of New York Press. Pp146.

Cicero, Marcus Tullius (1985 reprint) On the Nature of the Gods, 2.38. Penguin. Pp288.

Claxton, Guy (1994) Involuntary Simplicity: Changing Dysfunctional Habits of Consumption. Environmental Values 94 (3):72-78.

Clift, Wallis B. (1983) Jung and Christianity: The Challenge of Reconciliation. Dove Communications, Blackburn. Pp169.

Conze, Edward (1975) Further Buddhist Studies: Selected Essays by Edward Conze. Bruno Cassirer, Oxford.

Cornford, F.M. (1939) Plato and Parmenides: Parmenides' Way of Truth and Plato's Parmenides; Parmenides Fragment 2. Translated with Commentary. London.

Crosson, P. and Anderson, Jock (1993) Natural Resource Management Research Policy: Guidelines for Implementation in NARS. International Service for National Agricultural Research, The Hague.

Cunningham, Lawrence (1992) Thomas Merton: Spiritual Master: The Essential Writings. Contemplation in a World of Action. Paulist Press. Pp437.

Cupitt, Don (1995) The Last Philosophy. MCM Press, London. Pp149.

Cupitt, Don (2000) Millennial Dreams #6. The Spirit of Things, Radio National Sunday 6 February 2000 Interview by Rachael Kohn. <http://www.abc.net.au/rn/relig/spirit/stories/s98010.htm>

Curtin, Deanne (1995) Making Peace with the Earth: Indigenous Agriculture in the Green Revolution. Environmental Ethics 17:59-74.

Darwin, Charles (1874) The Descent of Man and Selection in Relation to Sex. Penguin reprint (2004) Pp791.

Dawkins, Richard (1990) The Selfish Gene. Oxford University Press. Pp368.

de Silva, Padmasiri (1990) Buddhist Environmental Ethics. Chapter in Badiner, Allan (editor) Dharma Gaia: A Harvest of Essays in Buddhism and Ecology. Parallax Press, Berkeley.

de Silva, Padmasiri (1998) Environmental Philosophy and Ethics in Buddhism. Macmillan Press, London. Pp195.

de Waal, Frans (1997) Good natured: The Origins of Right and Wrong in Humans and Other Animals. Harvard University Press, Cambridge, Massachusetts. Pp368.

DeNicolas, Antonio (1976) Meditations Through the Rg Veda. Nicolas Hays, New York.

d'Entreves, A.P. (1948). Aquinas, Selected Political Writings. (Translated by J.G. Dawson) Clarendon Press, Oxford.

Descartes, René (1901) The Method, Meditations and Philosophy of Descartes. Veitch, John (editor). Everymans Library, New York. Pp300.

Dewey, John (1959) Introduction to Lamont, Corliss. The Illusion of Immortality. Philosophical Library, New York. Pp303.

Dobson, Andrew (1995) Green Political Thought. Routledge, London.

Dryzek, John (1997) The Politics of Earth: Environmental Discourses. Oxford University Press, New York. Pp123.

Dubos, Rene (1972) A God Within. Scribner, New York. Pp325.

Dutch Government (1995) Commission on Ethical Aspects of Biotechnology in Plants. Chapter in Agriculture and Spirituality: Inter(agri)cultural Dialogue: Essay from the Crossroads Conference at Wageningen Agricultural University. International Books, Utretch. Pp141.

Duyvendak, J.J.L. (1993) Tao Te Ching by Lao Zi (Wisdom of the East Series). Turtle Publishing and John Murray. Pp172.

Earth Charter (2000) <http://www.earthcharter.org>

Eckel, Malcom (1997) Is There a Buddhist Philosophy of Nature? Chapter in Buddhism and Ecology: The Interconnection of Dharma and Deeds. Tucker, Mary and Williams, Duncan (editors). Harvard University Center for the Study of World Religions, Cambridge, Massachusetts. Pp467.

Edgerton, Franklin (1944) The Bhagavad Gita. Harper Torchbooks, New York.

Einstein, Albert (1996) The World as I See It. Collected and edited by Alice Calaprice. Princeton University Press, Princeton.

Eller, Cynthia (1992) The Impact of Christianity on Buddhist Non-Violence in the West. Chapter in Kraft, Kenneth (editor) Inner Peace, World Peace: Essays on Buddhism and Non-Violence. State University of New York Press, Albany.

Ellis, Robert (2001) How Buddhist was Plato? Western Buddhist Review Volume 3
< http://www.westernbuddhistreview.com/vol3/index.html >

Elton, Charles (1927) Animal Ecology. Press reprint (2001). Pp296.

Elwes, R. H. (1989) Ethics Including the Improvement of the Understanding: Benedict de Spinoza. Pp305.

Emerson, Ralph Waldo (1900) The Complete Prose Works of Ralph Waldo Emerson. Ward, Lock and Co., London. Pp656.

Falvey, Lindsay (1988) Introduction to Working Animals. MPW Australia, Melbourne. Pp198.

Falvey, Lindsay (1999) The Future for Smallholder Dairying . Chapter 24 in Falvey, Lindsay and Chantalakhana, Charan (1999) Smallholder Dairying in the Tropics. International Livestock Research Institute CGIAR, Nairobi.

Falvey, Lindsay (2000) Thai Agriculture: Golden Cradle of Millennia. Kasetsart University Press, Bangkok. Pp490.

Falvey, Lindsay (2001) AgriDhamma: The Duty of Professional Agriculturists. Translation of a lecture to agriculturists by Buddhadasa, 25 March 1991. <www.iid.org>

Falvey, Lindsay (2002) The Buddha's Gospel: A Buddhist Interpretation of Jesus' Words. Institute for International Development, Adelaide. Pp108. <www.iid.org>

Falvey, Lindsay (2004) Sustainability: Elusive or Illusion? Wise Environmental Intervention. Institute for International Development, Adelaide. Pp245. <www.iid.org>

Fitzpatrick, Noel (2000) Ecologically Sustainable Development or Sustainable Development. Focus 114: 2-6.

Flanagan, Sabina (1989) Hildegard of Bingen, 1098-1179: A Visionary life. Routledge, London. Pp230.

Flintoff, Everard (1980) Pyrrho and India. Phronesis: A Journal for Ancient Philosophy XXV (I): 88-107.

Fox, Stephen (1985) The American Conservation Movement: John Muir and His Legacy. Madison, Wisconsin.

Freud, Sigmund (1924) Thoughts on War and Death. Volume IV Collected Papers. International Psycho-Analytical Press. Pp305.

Fromm, Erich (1969) Escape From Freedom. Avon Books, New York. Pp333.

Fukuoka, Masanopu (1978) The One Straw Revolution: An Introduction to Natural Farming. Rodale Press, Emmaus.Pp181.

Fukuoka, Masanopu (1987) The Road back to Nature: Regaining the Paradise Lost. Japan Publications, Tokyo.Pp377.

Gabaude, Louis (1994) Buddhadasa's Contribution as a Human Being, as a Thai, as a Buddhist. Chapter in Sivaraksa, Sulak (editor) The Quest for a Just Society: The Legacy and Challenge of Buddhadasa Bhikku. Thai Inter-religious Commission for Development, Santi Pracha Dhamma Institute, Bangkok.

Gandhi, M.K. (1959) My Socialism. Navajivan Publishing House, Ahmedabad.

Gandhi, M.K. (1966) The Village Reconstruction. Bharatiya Vidya Bhavan, Bombay.

Geering, Lloyd (2001) Interview by Rachel Kohn on Radio National (Australia) Sunday 4 March 2001. <http://www.abc.net.au/rn/relig/spirit/stories/s253975.htm>

Gerson, Lloyd P. (1996) The Cambridge companion to Plotinus. Cambridge University Press, Cambridge.

Gerstenfeld, Manfred (1998) Judaism, Environmentalism and the Environment: Mapping and Analysis. The Jerusalem Insitute for Israel Studies and Rubin Mass, Jerusalem. Pp244.

Giri, J. (1984) quoted in WCED (1987) Our Common Future. World Commission on Environment and Development, Oxford University Press, Melbourne. Pp444.

Gomez, Luis (1992) Non-Violence and the Self in Early Buddhism. Chapter in Kraft, Kenneth (editor) Inner Peace, World Peace: Essays on Buddhism and Non-Violence. State University of New York Press, Albany.

Goss, Rita (1997) Buddhist Resources for Issues of Population, Consumption, and the Environment. Chapter in Tucker, Mary and Williams, Duncan (editors) Buddhism and Ecology: The Interconnection of Dharma and Deeds. Harvard University Center for the Study of World Religions, Cambridge, Massachusetts. Pp467.

Gordis, Robert (1986) Judaic Ethics for a Lawless World. The Jewish Theological Seminary of America, New York. Pp194
Gosling, David (2001) Religion and Ecology in India and Southeast Asia. Oxford University Press, London. Pp210.
Greenfield, Paul (2000) Wrap-Up Session. Paper presented to the Symposium of the Academy of Technological Sciences and Engineering, Brisbane, November 21-22.
Gregory, John (1991) The Neo-Platonists, Kyle Cathie.
Griffiths, Bede (1992) A New Vision of Reality: Western Science, Eastern Mysticism and Christian Faith. Harper and Collins, London, Pp304.
Gross, Rita (1997) Buddhist Resources for Issues of Population, Consumption, and the Environment. In Buddhism and Ecology: The Interconnection of Dharma and Deeds. Edited by Mary Tucker and Duncan Williams. Harvard University Center for the Study of World Religions, Cambridge, Massachusetts. Pp467.
Habel, Norman and Balabansky, Vicky (2001) The Earth Story in the New Testament: The Earth Bible Volume 5. Pilgrim Press. Pp225.
Hadas, Moses (1960) Essential Works of Stoicism. Bantam Books, New York.
Haeckel, Ernst (1866) General Morphology of Organisms. 2 vols. Reimer, Berlin.
Hallifax, Joan (1990) The Third Body: Buddhism, Shamanism, and Deep Ecology. Chapter in Badiner, Alan (editor) Dharma Gaia: A Harvest of Essays in Buddhism and Ecology. Parallax Press, Berkeley. Pp265.
Han, Beatrice (2000) Nietzsche and the Masters of Truth: The Pre-Socratics and Christ. Chapter in Lippitt, John and Urpeth, Jim (editors) Nietzsche and the Divine, Clinamen Press, Manchester.
Happold, F.C. (1991) Mysticism: A Study and an Anthology. Penguin. Pp416.
Hardin, Garrett (1968) The Tragedy of the Commons. Science 162:1243.
Hargrove, Eugene (1989) Aesthetics and Scientific Attitudes. Chapter 3 in Foundations of Environmental Ethics. Prentice Hall, New Jersey.
Harris, Elizabeth (1994) Violence and Disruption in Society: A Study of the Early Buddhist Texts. Buddhist Publication Society, Kandy, Sri Lanka.
Harris, Graham (2000) Is Australia Sustainable? Symposium of the Academy of Technological Sciences and Engineering, Brisbane, November 21-22.
Harris, Ian (1997) Buddhism and the Discourse of Environmental Concern: Some Methodological Problems Considered. Chapter in Buddhism and Ecology: The Interconnection of Dharma and Deeds. Tucker, Mary and Williams, Duncan (editors) Harvard University Center for the Study of World Religions, Cambridge, Massachusetts. Pp467.
Harrison, Paul (1999) Elements of Panthcism: Understanding the Divinity in Nature and the Universe. Element Books, United Kingdom Pp131.
Harrison, Peter (2001) Thc Bible, Protestantism, and the Rise of Natural Science. Cambridge University Press, Cambridge. Pp326.

Harrison, Ruth (1964) Animal Machines: The New Factory Farming Industry. Vincent Stuart, London.

Hatfield, J. L. and Karlen, D. L. (1993) Sustainable Agricultural Systems. Louis Publishers, Boca Raton, Florida, Pp307.

Hayward, Jeremy (1990) Ecology and the Experience of Sacredness. Chapter in Badiner, Alan (editor) Dharma Gaia: A Harvest of Essays in Buddhism and Ecology. Parallax Press, Berkeley. Pp265.

Hazell, Peter (2000) Agricultural Growth, Poverty Alleviation, and Environmental Sustainability: Having It All. 2020 Brief No. 59. International Food Policy Research Institute, Washington DC.

Helms, Douglas (1984) Walter Lowdermilk's Journey: From Forester to Land Conservationist. Environmental Review 8:133-145.

Hendrick, George (1977) Henry S. Salt, the Late Victorian Socialist and Thoreau. New England Quarterly 50:409-422.

Hillman, James (1995) A Psyche of the Size of the Earth: A Psychological Forward. In Roszak, Theodore (editor) Ecopsychology. Sierra Book Club, San Francisco. Page xxii.

Hodges, Jeff (1986) Harvesting the Suburbs: a Natural Approach to Backyard Gardening. Nature and Health Books, Dee Why West, Australia. Pp.139.

Holmgren, David (2002) Permaculture: Principles and Pathways Beyond Sustainability. Holmgren Design Services. <www.holmgren.com.au>

Howard, Robert (1943) An Agricultural Testament. Oxford University Press, London.

Hughes, Donald and Swan, Gemma (1986) How Much of the Earth is Sacred Space? Environmental Review 10:247-259.

Hughes, James (1993) Beginnings and Endings: The Buddhist Mythos of the Arising and Passing Away of the World. Chapter in Sivaraksa, Sulak (editor) Buddhist Perception for Desirable Societies in the Future: Papers prepared for the United Nations University. Thai Inter-religious Commission for Development and Sathirakoses-Nagapradipa Foundation, Bangkok.

Huizer, Gerrit (1995) Indigenous Knowledge and Popular Spirituality: A Challenge to Developmentalists. Chapter in Agriculture and Spirituality: Inter(agri)cultural Dialogue: Essay from the Crossroads Conference at Wageningen Agricultural University. International Books, Utretch. Pp141.

Hull, Brooks and Bold, Frederick (1994) Hell, Religion, and Cultural Change. Journal of Institutional and Theoretical Economics 150(3):447-64.

Hultzsch, E. (1925) Corpus Insriptionum Indicarum Volume 1: Inscriptions of Asoka (New Edition). Government of India/Clarendon Press, Oxford. Pp257.

Hume, Robert Ernest (1931) The Thirteen Brhadaranyaka Upanisads. Oxford University Press, Oxford. Section III:9.28.

Humphreys, Lee (1971) Pitfalls and Promises of Biblical Texts as a Basis For a Theology of Nature. Chapter in Stone, Glenn (editor) A New Ethic for a New Earth: Faith-Man-Nature Papers, 2. Handover, Connecticut.

Huxley, Aldous (1947) The Perennial Philosophy. Chato and Windus, London. Pp358.

Ingram, Paul (1990) Nature's Jewelled Net: Kukai's Ecological Buddhism. Pacific World 6:50-64

James, William (1994) The Varieties of Religious Experience: A Study in Human Nature: Being the Gifford Lectures on Natural Religion Delivered at Edinburgh in 1901-1902. Modern Library. Pp608.

Jefferies, Richard (1991) The Story of My Heart. eEdition, Amazon Press.

Johansson, Rune (1969) The Psychology of Nirvana. Allen & Unwin, London. Pp141.

Jorgenson, John (1989) Sensibility of the Insensible: The Genealogy of a Ch'an Aesthetic and the Passionate Dream of Poetic Creation. PhD Thesis, Australian National University, Canberra.

Kabilsingh, Chatsumarn (1990) Early Buddhist Views on Nature. Chapter in Badiner, Alan (editor) Dharma Gaia: A Harvest of Essays in Buddhism and Ecology. Parallax Press, Berkeley. Pp265.

Kant, Immanuel (1724-1804) Critique of Practical Reason. Book 2. Beck, Lewis White (translator). Maxwell Macmillan International, (c1993), New York.

Kaosa-ard, Mingsarn and Rerkasem, Benjaven (2000) The Growth and Sustainability of Agriculture in Asia. Oxford University Press, China. Pp303.

Kaza, Stephanie (1999) Overcoming the Grip of Consumerism. In Socially Engaged Buddhism for the New Millennium: Essays in Honor of Venerable Phra Dhammapitaka (Bhikkhu P.A. Payutto) on his 60th Birthday Anniversary. The Sathirakoses-Nagapradipa Foundation, Bangkok. Pp536.

Keown, Damien (1995) Buddhism and Bioethics. St. Martin's Press. London. Pp207.

Kern, H. (1943) The Jatakamala. Harvard Oriental Series, Volume 1. Harvard University Press, Cambridge, Massachusetts.

Kirsopp, Lake (1922) Immortality and the Modern Mind. Harvard University Press.

Kraft, Kenneth (1992) Introduction. In Inner Peace, World Peace: Essays on Buddhism and Non-Violence. State University of New York Press, Albany.

Kraft, Kenneth (1997) Nuclear Ecology and Engaged Buddhism. Chapter in Buddhism and Ecology: The Interconnection of Dharma and Deeds. Tucker, Mary and Williams, Duncan (editors). Harvard University Center for the Study of World Religions, Cambridge, Massachusetts. Pp467.

Krutch, Joseph (1978) The Great Chain of Life. Houghton Mifflin. Pp227.

Kulananda (1997) Western Buddhism: New Insights into the West's Fastest Growing Religion. Harper and Collins, London.

LaFleur, William R. (2000) Saving the Rainforest of Ethics: Society, Urgency, and the Study of Asia. Journal of Buddhist Ethics 7 (2000):169–181.

Lamont, Corliss (1959) The Illusion of Immortality. Philosophical Library, New York. Pp303.

Lan, Fung-Yu (1952) A History of Chinese Philosophy, translated by Derk Bodde. Princeton University Press, New Jersey.

Lancaster, Lewis (1997) Buddhism and Ecology: Collective Cultural Perceptions. Chapter in Buddhism and Ecology: The Interconnection of Dharma and Deeds. Tucker, Mary and Williams, Duncan (editors). Harvard University Center for the Study of World Religions, Cambridge, Massachusetts. Pp467.

Lawson-Tancred, Hugh (1986) De anima: On the Soul / Aristotle. Penguin, New York. Pp253.

Leake, John (2000) Aspects of Participatory Planning with Opium Growers in Vietnam. Master of Agricultural Science, University of Melbourne, Australia. Pp88.

Lebel, Louis and Steffen, Will (1999) Global Environmental Change and Sustainable Development in Southeast Asia: Science Plan for a SARCS Integrated Study. Southeast Asian Regional Committee for START, Bangkok. Pp 139.

Legge, James (1962) The Texts of Taoism, Dover Publications, New York.

Leopold, Aldo (1949) A Sand Country Almanac: Outdoor Essays and Reflections. Ballantine (reprint 1966 and 1990). Pp320.

Levins, Richard (1986) Science and Progress: Seven Developmentalist Myths in Agriculture. Monthly Review 38(3):13-20.

Levitt, Heidi (1999) The Development of Wisdom: An Analysis of Tibetan Buddhist Experience. The Journal of Humanistic Psychology, Beverly Hills. 39(2):86-105.

Locke, John (1690) Second Treatise of Government. C.Macpherson (editor). Hackett, Indianapolis reprint (1980). Pp24.

Lovelock, James (1982) A New Look at Life on Earth. Oxford University Press, New York.

Lovelock, James (1988) Ages of Gaia. Norton, New York.

Lovelock, James (2001) Homage to Gaia. Oxford University Press, London. Pp428.

Loy, David (1999) The Spiritual Roots of Modernity: Buddhist Reflections on the Idolatry of the Nation-State, Corporate Capitalism and Mechanistic Science. Chapter in Socially Engaged Buddhism for the New Millennium: Essays in Honor of Venerable Phra Dhammapitaka (Bhikkhu P.A. Payutto) on his 60th Birthday Anniversary. The Sathirakoses-Nagapradipa Foundation and The Foundation for Children, Bangkok. Pp536.

Luther, Martin (1484) Glossa Ordinaria, Venice.

Lynam, J. and Herdt, Robert (1989) Sense and Sustainability: Sustainability as an Objective in International Agricultural Research. Agricultural Economics 3(4):281-298.

Macy, Joanna (1983) Dharma and Development. Kumarian Press, West Hartford, Connecticut.

Malone, T F (1994) Sustainable Human Development: A Paradigm for the 21st Century - Challenge and Opportunity for Higher Education. A White Paper for the National Association of State Universities and Land Grant Colleges; Commission on Food, Environment and Renewal Resources, Washington DC.

Malthus, Thomas (1914) An Essay on Population. Dent, London.

Marietta, Don Jr (1984) Environmentalism, Feminism and the Future of American Society. The Humanist 44:15-30.

Marsh, George Perkins (1958) Versatile Vermonter. Columbia University Press, New York.

Matilal, Bimal Krishna (1981) The Central Philosophy of Jainsim. L.D. Institute of Indology, Ahmedabad, India.

Meadows, Donella, Meadows, Dennis, Randers, Jorgen and Behrens, William (1974) The Limits to Growth: A Report for the Club of Rome's Project on the Predicament of Mankind. Universe Books, New York. Pp205.

Miles, Jack (1997) God, A Biography. Alfred A. Knopf, New York.Pp.403.

Moltmann, Jurgen (1985) God in Creation: An Ecological Doctrine. The Gifford Lectures 1984-1985. Translated by Margaret Kohl. SCM Press, London. Pp365.

Moore, Thomas (2000) Placebos, Faith, and Morals: Or Why Religion. Hoover Institution, Stanford University.
<http://www.stanford.edu/~moore/Placebos.html>

Munasinghe, Mohan (1993) Environmental Economics and Sustainable Development. World Bank Environment Paper No. 3, World Bank, Washington DC.

Naess, Arne (1973) The Shadow and the Deep, Long-Range Ecology Movement: A Summary.

Naess, Arne (1998) Beautiful Action: Its Function in the Ecological Crisis. Environmental Values 2.

Nash, Roderick (1989) The Rights of Nature: A History of Environmental Ethics. The University of Wisconsin Press, Madison. Pp 290.

Nicklin, Donald (2000) Sustainability - The Issues and the Obligations. Paper presented to the Symposium of the Academy of Technological Sciences and Engineering, Brisbane, November 21-22.

Norberg-Hodge, Helena (1999) Buddhist Engagement in the Global Economy. Chapter in Socially Engaged Buddhism for the New Millennium: Essays in Honor of Venerable Phra Dhammapitaka (Bhikkhu P.A. Payutto) on his 60th Birthday Anniversary. The Sathirakoses-Nagapradipa Foundation and The Foundation for Children, Bangkok. Pp536.

Nyanasamvara, Somdet Phra (2001) Supreme Patriarch of Thailand, quoted in Seeds of Peace 18(1):52.

O'Brien, Brian (2000) Risking Australia's Future with the Precautionary Principle. Focus 114:16- 21.
O'Donohue, John (1998) Eternal Echoes: Exploring Our Hunger to Belong. Bantam, New York. Pp389.
Open Letter (2000) Open Letter from Leading Scientists and Religious Leaders. Global Forum Conference, Moscow <http://earthrenewal.org/Open_letter_to_the_religious_.htm>
Orr, D. W. (1992) Ecological Literacy: Education and Transition to a Post Modern World. State University of New York Press, New York.
Palmer, Joy (2001) Fifty Key Thinkers on the Environment. Routledge, London
Passmore, John (1974) Man's Responsibility for Nature: Ecological Problems and Western Traditions. Scribner, New York. Pp213.
Passmore, John (1975) The Treatment of Animals. Journal of the History of Ideas 36:195-218
Payutto, P.A. (1993) Toward Sustainable Science: A Buddhist Look at Trends in Scientific Development. Buddhadhamma Foundation, Bangkok. Pp175.
Payutto, P.A. (1994) Acceptance Speech of Phra Payutto for the 1994 UNESCO Peace Education Prize. Chapter in Peace Through Freedom and Happiness. Bangkok.
Payutto, P.A. (1995) Buddhadhamma: Natural Laws and Values for Life. Translated by Grant A. Olson. State University of New York Press, Albany. Pp302.
Payutto, P.A. (2000) A Constitution for Living: Buddhist Principles for a Fruitful and Harmonious Life. [Thai: Phra Thammapidok (2543) Thamanun Chewit.] Translated by Bruce Evans. 26th Edition. Kasetsart University, Bangkok. Pp 85.
Payutto, P.A..(1990) Buddhist Education. Chapter in Sivaraksa, Sulak (editor) Radical Conservatism: Buddhism in the Contemporary World: Articles in Honour of Bhikkhu Buddhadasa's 84th Birthday Anniversary. Thai Inter-religious Commmission - International Network of Engaged Buddhists. Pp558.
Payutto, P.A. (1993) Good, Evil and Beyond: Karma in the Buddha's Teaching. Translated by Bhikkhu Puriso. Buddhadhamma Foundation Publications, Bangkok.
Payutto, P.A. (2001) The Middle Teaching (Majjhena Dhammadesano.) New translation of sections of Buddhadamma by Bruce Evans (unpublished manuscript).
Peacock, James (2000) Sustainable Agriculture. Symposium of the Academy of Technological Sciences and Engineering, Brisbane, November 21-22.
Peck, Scott (2001) A World Waiting to be Born: Civility Rediscovered. Bantam Books, New York. Pp366.
Petersson, Torsten (1920) Cicero: A Biography. University of California, Berkeley. Pp699.

Plank, Max (1984) The Mystery of Our Being. Chapter in Wilbur, Ken (editor) Quantum Questions New Science Library, Boston. Pp153.
Pope, Alexander (1733) Essay on Man. Dover reprint (1994) Pp99.
Prabhavananda, Swami and Isherwood, Christopher (1987) The Song of God: Bhagavad Gita. Vedanta Press.
Prasety, Teddy (1993) Buddhism and the Evolution of Society. Chapter in Sivaraksa, Sulak (editor) Buddhist Perception for Desirable Societies in the Future: Papers prepared for the United Nations University. Thai Inter-religious Commission for Development. Sathirakoses-Nagapradipa Foundation, Bangkok.
Pryor, Frederick (1990) A Buddhist Economic System - in Principle. American Journal of Economics and Sociology: 49(3):339-351.
Pryor, Frederick (1991) A Buddhist Economic System - in Practice. American Journal of Economics and Sociology. 50(1):17-32.
Queen, Christopher (1996) Introduction. Chapter in Queen, Christopher and King, Sallie (editors) Engaged Buddhism: Buddhist Liberation Movements in Asia. State University of New York Press, Albany. Pp115.
Rahula, Walpola Sri (1972) What the Buddha Taught (Revised Edition). Gordon Fraser, UK. Pp151.
Ramakrishnan, P.S. (1992) Shifting Agriculture and Sustainable Development: An Interdisciplinary Study from North-Eastern India. UNESCO and the Parthenon Publishing Group, New Jersey. Pp424.
Read, Herbert and Hull, R.F.C. (1970) The Collected Works of C.G. Jung, Princeton University Press, Princeton.
Reanney, Darryl (1991) The Death of Forever: A New Future of Human Consciousness. Longnan Cheshire, Pp225.
Reichelt, Russell (2000) Sustainable Management of the Great Barrier Reef. Paper presented to the Symposium of the Academy of Technological Sciences and Engineering, Brisbane, November 21-22.
Revel, Jean-François and Ricard, Matthieu (1998) The Monk and the Philosopher: East Meets West in a Father-Son Dialogue. Thorsorns, London.
Ritchie, Donald (1971) The Island Sea. Weatherhill, Tokyo.
Ross, W.D. (1924) Aristotle's Metaphysics: A Revised Text. Clarendon Press, Oxford. Pp983.
RPDB (1997) Concepts of His Majesty the King on Development. Royal Projects Development Board, Bangkok.
Ryle, Gilbert (1966) Plato's Progress. Cambridge University Press, Cambridge.
Saarmati (2000) Green Buddhism and the Hierarchy of Compassion. (Alan Sponberg). Western Buddhist Review Volume 1. <http://www.westernbuddhistreview.com/vol2/ecological_self.html>
Saarmati (2001) The Buddhist Conception of an Ecological Self. (Alan Sponberg). Western Buddhist Review Volume 2. <http://www.westernbuddhistreview.com/vol2/ecological_self.html>

Sabasri, Sanga (1994) Bioresources and Sustainable Use in Southeast Asia. Annual Oration to the Australian Academy of Technological Sciences and Engineering. <http://www.atse.org.au>

Salt, Henry (1921) Seventy Years Among Savages. Allen and Unwin, London. Pp251.

Sangarakshita (1986) Alternative Traditions. Windhorse, Glasgow. Pp197.

Sangharakshita (1994) The FWBO and Protestant Buddhism: An Affirmation and a Protest. Windhorse Publications, Glasgow. Pp187.

Santha-Anand, Suwanna (1994) Buddhadasa's Bitter Potions as Legacies for Thai Society. Chapter in Sivaraksa, Sulak (editor) The Quest for a Just Society: The Legacy and Challenge of Buddhadasa Bhikku. Thai Inter-religious Commission for Development, Santi Pracha Dhamma Institute, Bangkok.

Santikaro, Bhikkhu (1996) Buddhadasa Bhikkhu: Life and Society through the Natural Eyes of Voidness. Chapter in Queen, Christopher and King, Sallie (editor) Engaged Buddhism: Buddhist Liberation Movements in Asia. State University of New York Press, Albany. Pp446.

Santmire, Paul (1985) The Travail of Nature: The Ambiguous Ecological Promise of Christian Theology. Fortress Press. Pp274.

Sayadaw, Narada Mula Patthana (1969, 1981) Conditional Relations (Patthana). Pali Text Society, London. 2 Volumes.

Schilthuis, Willy (1994) Biodynamic Agriculture: Rudolf Steiner's Ideas in Practice. Floris Books, Edinburgh. Pp111.

Schmithausen, Lambert (1991) Buddhism and Nature. The International Institute for Buddhist Studies, Tokyo.

Schumacher, E.F. (1973) Small is Beautiful: A Study of Economics as if People Mattered. Blonde Briggs, London.

Schweitzer, Albert (1950) The Animal World of Albert Schweitzer. Translated by Charles Joy. Boston.

Seccombe, John (2000) Sustainability of the Great Artesian Basin. Paper presented to the Symposium of the Academy of Technological Sciences and Engineering, Brisbane, November 21-22.

Sessions, George (1977) Spinoza and Jeffers on Men in Nature. Inquiry 20:481-528.

Shiva, Vandana (1995) Nature, Creativity and the Arrogance of Patenting Life-forms. Chapter in Agriculture and Spirituality: Inter(agri)cultural Dialogue: Essay from the Crossroads Conference at Wageningen Agricultural University. International Books, Utretch. Pp141.

Shivakoti, Ganesh, Varughese, George, Ostrom, Elinor, Shukla, Ashutosh, and Tajpa, Ganesh (1997) People and Participation in Sustainable Development: Understanding the Dynamics of Natural Resource Systems. Proceedings on an International Conference, Institute of Agricultural and Animal Science, Tribhuvan University, Nepal, March 17-21 1996.

Simon, Herbert (1990) A Mechanism for Social Selection and Successful Altruism. Science 250:1665-1668.
Singer, Peter (1975) Animal Liberation: New Ethics for our Treatment of Animals. Random House, New York. Pp301.
Singer, Peter (1993) How Are We To Live? Ethics in an Age of Self-Interest. Mandarin-Prometheus, Melbourne. Pp262.
Singer, Peter (2000) Writings on an Ethical Life. Harper & Collins, New York. Pp.361.
Singer, Peter (2001) Animal Liberation. Ecco. Pp352.
Sittler, Joseph (1954) A Theology For Earth. Christian Scholar 37:367-374.
Sivaraksa, Sulak (1992) Buddhism and Contemporary International Trends. Chapter in Kraft, Kenneth (editor) Inner Peace, World Peace: Essays on Buddhism and Non-Violence. State University of New York Press, Albany.
Sivaraksa, Sulak (1993) Radical Conservatism. Buddhism in the Contemporary World: Articles in Honour of Bhikkhu Buddhadasa's 84th Birthday Anniversary. Thai Inter-religious Commission for Development and International Network of Engaged Buddhists. Pp558.
Snyder, Gary (1974) Turtle Island. New Directions, New York Pp114.
Spanel, Ann (1988) Interview with Vandana Shiva. Woman of Power, Volume 9.
Sponberg, Alan (1997) Green Buddhism and the Hierarchy of Compassion. Chapter in Tucker, Mary and Williams, Duncan (editors) Buddhism and Ecology: The Interconnection of Dharma and Deeds., Harvard University Center for the Study of World Religions, Cambridge, Massachusetts. Pp467.
Spong, John Shelby (1998) The Spirit of Things. Radio National Interview with Rachael Kohn on Sunday 16 August 1998. <http://www.abc.net.au/rn/relig/spirit/stories/>
Steele, Zelma (1942) Angel in Top Hat. Harper, New York.
Stefferud, Alfred (1968) Christians and the Good Earth. Faith-Man-Nature. Alexandria, Virginia.
Steiner (1974) Agriculture : A Course of Eight Lectures; English translation by George Adams; preface by Ehrenfried Pfeiffer. Bio-Dynamic Agricultural Association, London.
Stone, Christopher (1974) Should Trees Have Standing? Toward Legal Rights for Natural Objects. Los Altos, California.
Straughan, Roger (1992) Ethics, Morality and Crop Biotechnology. ICI Seeds, Reading University, England.
Subuti (1995) Sangarakshita: A New Voice in the Buddhist Tradition. Windhorse Publications, Birmingham. Pp324.
Sueki, Fimihiko (1994) Annen: The Philosopher Who Japanised Buddhism. Acta Asiatica 66.
Suzuki, D.T. (1932) The Lankavatara Sutra. Routledge and Kegan Paul, London.

Swarminathan, M. S. (1994) Report on Presentation to the International Conference on Population and Development held in Cairo in 1994. CERES 27:1:11.
Swearer, Donald (1991) Me and Mine: Selected Essays of Buddhadasa. Sri Satguru Publications, Indian Books Centre, Delhi. Pp235.
Swearer, Donald (1992) Exemplars of Non-Violence in Buddhism. Chapter in Kraft, Kenneth (editor) Inner Peace, World Peace: Essays on Buddhism and Non-Violence. State University of New York Press, Albany.
Swearer, Donald (1994) Three Legacies of Bhikku Buddhadasa. Chapter in Sivaraksa, Sulak (editor) The Quest for a Just Society: The Legacy and Challenge of Buddhadasa Bhikku. Thai Inter-religious Commission for Development, Santi Pracha Dhamma Institute, Bangkok.
Tagore, Rabindra Nath (1941) Rabindra Natha Tagore on Rural Reconstruction. Government of India, New Delhi.
Tambiah, S.J. (1968) Dialectic in Practical Religion. Cambridge Papers in Social Anthropology 5, In Leach, E.R. (editor) Cambridge University Press, Cambridge. Pp216.
Tambiah, S.J. (1976) World Conqueror and World Renouncer: A Study of Buddhism and Polity in Thailand Against a Historical Background. Cambridge University Press, New York. Pp557.
Tauler, Johannes (1300-1361) The Inner Way. Dominican.
Teilhard de Chardin, Pierre (1972) Le Milieu Divine. French and European Publications. Pp192.
Thompson, Paul (1995) The Spirit of the Soil: Agriculture and Environmental Ethics. Routledge, London. Pp36.
Thoreau, Henry David (1999) Walden (and Civil Disobedience) with a New Introduction by Merwin, W.S. Signet Classic, New York. Pp305.
Tillich, Paul (1949) The Shaking of the Foundations. Penguin, Middlesex. Pp.186.
Tillich, Paul (1973) Systematic Theology Volume 3. University of Chicago Press. Pp300.
Tillich, Paul (1976) Systematic Theology: Life and the Spirit: History and the Kingdom of God Volume 3. Pp434.
Timmerman, Peter (1992) It is Dark Outside: Western Buddhism from the Enlightenment to the Global Crisis. Chapter in Batchelor, Martine and Brown, Kerry (1992) Buddhism and Ecology. Cassell Publishers Ltd, London.
Tobias, Michael (1991) Life Force: The World of Jainism. Asian Humanities Press, Berkeley, California.
Torrey, Bradford (1906). The Writings of Henry Thoreau. Boston.
Traherne, Thomas (1988) Centuries of Meditation. Bridge Logos After Faith /Clarendon.

Tucker, Mary and Grim, John (1997) Series Forward. In, Buddhism and Ecology: The Interconnection of Dharma and Deeds. Tucker, Mary and Williams, Duncan (editors). Harvard University Center for the Study of World Religions, Cambridge, Massachusetts. Pp467.

Twersky, Isadore (1972) A Maimonides Reader. Behrman House, New York.

van Willenswaard, Hans (1999) Buddhism and Agriculture. Chapter in Socially Engaged Buddhism for the New Millennium: Essays in Honor of Venerable Phra Dhammapitaka (Bhikkhu P.A. Payutto) on his 60th Birthday Anniversary. The Sathirakoses-Nagapradipa Foundation, Bangkok. Pp536.

Verhoog, Henk (1995) The Oppressed Tradition of Caring Objectivity in Western Culture. Chapter in Agriculture and Spirituality: Inter(agri)cultural Dialogue: Essay from the Crossroads Conference at Wageningen Agricultural University. International Books, Utretch. Pp141.

Verkleij, Frans (1995) Spirituality and Ecological Agriculture. Chapter in Agriculture and Spirituality: Inter(agri)cultural Dialogue: Crossroads Conference at Wageningen Agricultural University. International Books, Utretch. Pp141.

Volk, Tyler (1997) Gaia's Body: Toward a Physiology of Earth. Copernicus. Pp269.

Wasi, Prawase (1989) Wikritmubanthai. Chapter in Chumchonphatthana Chababphiset. Muban Press. Pp109.

Wasi, Prawase (1992) Sasana Kap Singwaetlom. Singwaet Lom 35.

Wasi, Prawase (1993) Alternative Buddhist Agriculture. Chapter in Sivaraksa, Sulak (editor) Radical Conservatism. Buddhism in the Contemporary World: Articles in Honour of Bhikkhu Buddhadasa's 84th Birthday Anniversary. Thai Inter-religious Commission for Development. Pp.558.

Wasi, Prawase (1994) Khwam Lak Lai Thang Chiwaphap Kap Karn Phattana Yang Yangyun.

Wasi, Prawase (1995) Bottham Lok Si Khiew: Jit Samnuk Nai Khong Manut Chart. Sathan Karn Singwaetlom Thai 2538.

Wasi, Prawase (1996) Hon Thang Rod Sangkhon Thai. Phim Di Press, Bangkok Pp.31.

Wasi, Prawase (1998a) Buddhist Agriculture and the Tranquility of Thai Society. Chapter in Phongphit, Seripo and Bennoun, Robert (editors) Turning Point of Thai Farmers, Culture and Development Series 2. Thai Institute for Rural Development, Bangkok.

Wasi, Prawase (1998b) Suanmokh Thammakay Santiasok. Mochaoban Press, Bangkok. Pp.106.

Wasi, Prawase (1998c) Vrachakhomtambol: Yuthasastphua Setakitphophieng Silitham Le Sukhaphap, Bangkok. Pp48.

Wasi, Prawase (1999a) Sangkhom Di Mai Me Khai Yak Dai Tong Ruam Sang. Newspaper article photocopy from Index No I1831708X.

Wasi, Prawase (1999b) Setthakit Pho Phiang Le Prachasongkhom: Naew Thang Phlik Fuun Setthakit Songkhom. Mo Chao Ban Publishers, Bangkok. Pp.73

Waterfield, Robin (1996) Physics: Aristotle. Oxford University Press, Oxford. Pp301.

Watts, Don (2000) Intellectual Capability: The Foundation of a Sustainable Nation. Paper presented to the Symposium of the Academy of Technological Sciences and Engineering, Brisbane, November 21-22.

WCED (1987) Our Common Future. World Commission on Environment and Development, Oxford University Press, Melbourne. Pp444.

White, Lynn Jr (1967) The Historical Roots of Our Ecological Crisis. Science 155:1203-1207.

Whitehead, A.N. (1996) Religion in the Making: The Lowell Lectures. Fordham, New York.

Whitehead, Alfred North (1920) The Concept of Nature: Tarner Lectures. Cambridge University Press, Cambridge. Pp212.

Wilson, Edward (1984) Biofilia. Cambridge, Massachusetts.

Winkworth, Susanna (2004) The Theologia Germanica of Martin Luther. Dover Publications. Pp144.

Wood, Harold Jr. (1985) Modern Pantheism as an Approach to Environmental Ethics. Environmental Ethics 7:51-63.

Yuasa, Nobuyuki (1966) Basho: The Narrow Road to the Deep North (and Other Travel Sketches). Penguin, Baltimore.

Zhu ZL, Chen D (2002) Nitrogen fertilizer use in China - Contributions to food production, impacts on the environment and best management strategies. Nutrient Cycling in Agroecosystems 63(2): 117-127.

Zillman, John (2000) The State of Greenhouse Science. Paper presented to the Symposium of the Academy of Technological Sciences and Engineering, Brisbane, November 21-22.

Zweers, Wim (1995) Ecological Spirituality as Point of Departure fro an Intercultural Dialogue. Chapter in Agriculture and Spirituality: Inter(agri)cultural Dialogue: Essay from the Crossroads Conference at Wageningen Agricultural University. International Books, Utretch. Pp141.